Diseases of HERBACEOUS PERENNIALS

Mark L. Gleason, Margery L. Daughtrey,
Ann R. Chase, Gary W. Moorman,
and Daren S. Mueller

The American Phytopathological Society
St. Paul, Minnesota 55121 U.S.A.

Front cover image courtesy M. Daughtrey

Back cover images courtesy P. Ek (daylily and monarda) and M. Daughtrey

Table of Contents images courtesy M. Tobiasz (tulips) and M. Daughtrey

Interior designer: Mary Sailer

Compositor: Diana Van Winkle

This book was formatted from computer files submitted to The American Phytopathological Society by the authors of the volume. No editing or proofreading has been done by the publisher.

Reference in this publication to a trademark, proprietary product, or company name is intended for explicit description only and does not imply approval or recommendation to the exclusion of others that may be suitable.

Library of Congress Control Number: 2009925591

International Standard Book Number: 978-0-89054-374-0

Printed in the United States of America on acid-free paper

The American Phytopathological Society
3340 Pilot Knob Road
St. Paul, Minnesota 55121, U.S.A.

Preface

IN THE WORLD of ornamental production and gardening, beauty is foremost. There are literally hundreds of books that tell us how to grow the prettiest yarrow, or why we should buy a beautiful campanula, or regale us with the low maintenance value of that lovely coneflower.

So it would appear to be difficult to get excited about a book whose topic is to show how bad those same plants can look. It would seem to be a challenge to expect a reader to want to learn about the dreadful things that can assault their favorite perennial, and then expect the same reader to look at photos of that damage as well. A bit like viewing a car wreck, some would say. Not so! That my plants are going to get mugged occasionally by diseases is not in question. The real questions are—do I know how to identify, prevent, or reduce the damage?

To believe that all perennials are perfect is to believe that all men can fix cars. Pathogens, given the right environment, result in leaf spots, root damage, and death. These authors have been at the forefront of the pathogen wars for decades and their common sense approach to plant disease management is highly respected in scientific, commercial and garden circles. They provide more of the same in this book.

I approached the book as a producer, a landscaper, and gardener. I wanted to know what pathogens might attack my Acanthus, Echinacea and hundreds of other perennials that I just have to have. However, since I have no lab to evaluate leaf spots or root swellings, and I am not a plant pathologist, I also wanted good photos to verify problems I might see in my perennials. And, most important, I wanted some common sense ideas explaining how to reduce the incidence of these awful problems.

I have not been disappointed. For every plant, this book tells me what might occur, it shows me what these occurrences might look like, and best of all, provides sensible suggestions and scientifically proven solutions. I didn't want to hear about chemicals because it is well known that chemicals that are here today may be gone tomorrow—and I want to use as few chemicals as possible. I can easily understand and implement suggestions such as "choose a better cultivar, this one is highly susceptible" or "dead heading is the best prevention," or "simply provide wider spacing in the garden." Common sense runs in the veins of the authors, and the book confirms it.

Diseases may not be fashionable topics in this world of perennial beauty, but they are real, and this book tackles the reality head on. I am excited—I recommend it without question.

Dr. Allan Armitage
Professor of Horticulture
University of Georgia
Author: Herbaceous Perennial Plants, 3rd ed.

About the Authors

MARK GLEASON is a professor and extension plant pathologist in the Department of Plant Pathology at Iowa State University, where he has focused on diseases of horticultural commodities, including ornamentals, for the past 24 years.

MARGERY DAUGHTREY is a senior extension associate with Cornell University's Department of Plant Pathology and Plant-Microbe Biology. She has 30 years of experience in extension education and applied research on diseases of greenhouse and nursery crops at the Long Island Horticultural Research & Extension Center.

ANN CHASE is the founder and CEO of Chase Horticultural Research, a private research and diagnostic firm located in Mt. Aukum, CA. She was professor of Plant Pathology at the University of Florida Central Florida Research and Education Center in Apopka from 1979 to 1994.

GARY MOORMAN is a professor in the Pennsylvania State University Dept. of Plant Pathology, where he has worked in woody ornamental and floriculture pathology research, extension and teaching for the past 27 years. Previously he was a member of the Department of Plant Pathology of the University of Massachusetts at the Suburban Experiment Station in Waltham, MA working on diseases of vegetables and floricultural crops from 1978 to 1982.

DAREN MUELLER has been with the Department of Plant Pathology, Iowa State University since 2003, and now serves as an extension program specialist. He has expertise on white mold and rust diseases of ornamental and field crops, including two invasive introduced diseases: daylily rust and Asian soybean rust.

Acknowledgments

THE AUTHORS WOULD like to express special gratitude to Winnie Gleason, Robert Kent, Mike Zemke, Fran Moorman, and Katie Mueller for their patience and support during the writing and illustration-gathering for this book. We have visited many beautiful gardens and well-run nurseries to collect images of herbaceous perennials over the past several years, and this part of the work was indeed a pleasure. Your hospitality to our cameras has been deeply appreciated. Another special thank-you goes to Allan Armitage, whose excellent book, *Herbaceous Perennial Plants—A Treatise on Their Identification, Culture, and Garden Attributes*, now in its third edition, supplies such well-informed perspective on the growth of herbaceous perennials in different areas of the United States. Many other people also deserve special mention for their important contributions to this book. Janna Beckerman, Scott Clark, Stephanie Cohen, June Croon, Jim Glover, Caroline Kiang, Joseph LaForest, Ben Lockhart, Barry Menser, Janice Munson, Theo Overdevest, Leanne Pundt, Henriette Suhr, Ellen Talmage, Maria Tobiasz and Andre Viette, in particular, were amazingly helpful. A number of talented photographers who have shared their images with us are listed on page 255. This book couldn't have been done without a lot of help from our colleagues and associates—so thanks to you all!

SPONSOR

IOWA NURSERYMEN'S RESEARCH CORPORATION

We are especially grateful to the Iowa Nurserymen's Research Corporation, whose invaluable financial support made this book possible.

Contents

Diseases of HERBACEOUS PERENNIALS

Introduction

NORTH AMERICAN GARDENERS are planting a rapidly expanding range of herbaceous perennials. New plants are flooding the marketplace, and formerly obscure plants are becoming widely available. Learning how to take care of so many new kinds of plants can be a challenge.

Dealing with plant diseases is part of this challenge. Knowledge about diseases of herbaceous perennials is mushrooming almost as fast as the perennials industry, but is scattered among many plant pathologists, horticulturists, books, websites, extension publications, and scientific articles.

The goal of this book is to help you to recognize the major diseases of herbaceous perennial ornamentals and to manage them effectively. Our geographic focus is North America (the United States, Canada, and Mexico); diseases known elsewhere in the world will not necessarily be mentioned here. Because the ornamental plant trade today is a global industry, however, problems currently restricted to other continents may well be brought into North America in the future. Similarly, diseases known to occur on wild relatives of ornamentals may occasionally appear in the garden, so we have reported problems that occur on the host plant's genus even if they have not been common on garden plants.

This book is intended for home gardeners, commercial plant producers and retailers, landscape managers, diagnosticians, educators, and anyone else who loves perennials.

What is an herbaceous perennial?

We define herbaceous perennials as ornamental plants that lack woody tissue and are perennial (persist from year to year) in at least some part of North America. The book emphasizes perennials that are generally grown outdoors, rather than greenhouse or interiorscape species. Some of the plants covered are listed as "tropical perennials" because they are not frost-hardy.

How to use the book

This Introduction is followed by a section describing basic strategies for diagnosing and managing diseases of herbaceous perennials. It explains why Integrated Pest Management is important, and outlines some helpful practices to keep diseases in check.

Next come short profiles of 12 major types of diseases that attack herbaceous perennials. This section will help readers understand how different types of disease-causing agents live their lives, and how best to protect plants against them.

The main section of the book is organized alphabetically by the genus of perennial plant. To help readers rapidly find information for a particular plant of interest, the genera are organized alphabetically. If you are unfamiliar with Latin names of the plants you are interested in, check the Common Name/Latin Name Index on page 249 to find the Latin name of the plant you are seeking. Photos of some of the major diseases of each genus are combined with text descriptions.

We have included as many genera of herbaceous perennials as possible. You may find that some plants that you grow do not appear in the book; this may be because because no diseases have been reported on these plants from North America.

Diagnosing and Managing Diseases of Herbaceous Perennials

SINCE THIS BOOK focuses on the recognition and management of plant diseases, a few key ideas from the science of plant pathology will be helpful to the reader. Symptoms on plants may be caused by either non-contagious factors (such as cultural errors in watering or fertilization) or by microorganisms, which are contagious. Every contagious plant disease involves an interaction between three players: the host plant (the victim), the pathogen (the attacker), and the environment. Plants get sick only when a pathogen comes in contact with a plant that is susceptible to it, and then only when the environment allows the pathogen to attack.

Even though there are thousands of different species of plant pathogens—fungi, bacteria, nematodes, viruses, and even parasitic plants—deciding how to manage them is simpler than it looks. Luckily, each type of host plant—hosta, daylily, astilbe, etc.—is attacked by a relatively short list of pathogens. So once you identify your host plant, you have already taken a huge step toward making the diagnosis by reducing the number of possible diseases.

Another plus is that pathogens come in only a few types. The following section of this book acquaints you with 12 major pathogen types, and provides tips on managing each type. For example, although many species of powdery mildew fungi attack perennials, almost all powdery mildew species behave similarly. So once you understand how to manage powdery mildew on dahlia, the same general IPM approach is likely to work on delphinium—even though the species of powdery mildew fungi usually differ on plants from different plant families. Learning the basic attack strategy used by each type of pathogen will give you a head start in deciding how to prevent the diseases they cause.

Diagnosing the cause of a plant disease puts you on the road to managing it successfully. Once you know what is causing the symptoms on your plants, you can zero in on the best ways to control the disease—and take steps to avoid it in the future. This book is intended to help you diagnose disease problems by matching what you see on your plants with the book's photos and text descriptions. In the main body of the book, we have described the diseases of the most commonly planted genera of herbaceous perennials. Additional text and Internet resources that can help with diagnoses are listed at the end of the book.

Extension specialists and plant diagnostic clinics across the United States provide invaluable local and regional expertise in diagnosis as well as treatment and prevention advice. It's well worth your while to get acquainted with the perennial plant disease experts in your area, in order to learn how they can help you. But you don't have to be a plant scientist to be good at diagnosis. Armed with a few well-illustrated resources, anyone who loves plants and is a careful observer of clues can become an excellent disease troubleshooter.

Being a grower of herbaceous perennials, whether for sale, public display, or personal pleasure, means making many decisions that affect the risk of disease outbreaks. Which varieties you choose, the cultural practices you use to grow them, and which (if any) chemical or biological control products you apply can all affect the appearance of your plants and the cost of pest management.

Each decision a grower makes can influence the other decisions. For example, deciding to grow a variety that is very resistant to major diseases can cut the risk of having severe outbreaks of those diseases. On the other hand, growing a disease-prone variety because you love it, or because your buyers really want it, means that you may need to put extra effort into protecting it from diseases. On the cultural side, it may be simpler or less expensive to use sprinklers rather than drip irrigation. But sprinklers can raise the risk of fungal and bacterial diseases that thrive on wet foliage, so you may sometimes need pesticide sprays to protect the plants.

How do you decide what disease management strategies make the most sense for you? Making sensible choices seems challenging when you are growing many different species of herbaceous perennials, each with its own growth characteristics and risks. To avoid getting lost in the details, it's helpful to rely on some basic disease management principles.

Integrated Pest Management (IPM) is a strategy for bringing together the most appropriate management practices—including cultural, genetic resistance, biological, mechanical, and chemical options—to keep diseases, insects, weeds, and other pests in check, with minimal harm to the environment and your checkbook. An IPM plan should make sense for your particular environment, whether it is a commercial nursery, retail garden center, public landscape, or home garden. Likewise, disease management plans need to make sense alongside strategies for controlling insects and other pests, and for meeting the grower's goals. We have described IPM strategies for disease management throughout this book.

General Types of Diseases

Fungal Leaf Spots

Ascochyta blight on clematis.

INTRODUCTION

Fungal leaf spots are among the most commonly occurring diseases of herbaceous perennials. A multitude of fungi can cause leaf spot symptoms, including species of *Alternaria, Ascochyta, Cercospora, Corynespora, Cylindrocladium, Cylindrosporium, Didymella, Entyloma, Fabraea, Marssonina, Phyllosticta, Pleospora, Ramularia, Septoria,* and many others. Although some leaf spot fungi attack a wide range of perennials, many are specific to one or a few plant genera, species, or varieties. Leaf spot damage ranges from minor to severe depending on the interplay of fungus, host plant, and environmental conditions.

Symptoms

Depending on the fungus-host combination, leaf spots can vary in size, shape, and color. Some leaf spot fungi produce tiny, dark-colored fruiting bodies (spore-producing structures) in the spots and others develop distinctive-looking masses of spores. A hand lens can help you to spot these clues. Still other leaf spot fungi produce no conspicuous signs. When conditions are favorable, leaf spots may expand in size or number, merging to create larger dead areas referred to as leaf blight. In severe cases, all of the foliage may become blighted.

Ecology

Almost all leaf spot diseases thrive in warm, moist weather. To invade, spores of these fungi need leaf surfaces to remain wet for at least a minimum period of time—several to many hours—depending on the air temperature and the host plant. Prolonged spells of warm, rainy weather set the stage for severe leaf spot outbreaks. Most leaf spot fungi are adept at surviving unfavorable periods (winter or drought periods) burrowed inside dead

Heterosporium leaf spot on iris.

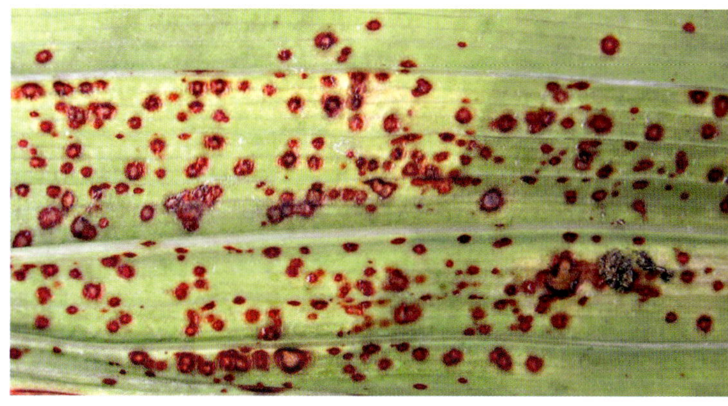

Stemphylium leaf spot on gladiolus.

leaf tissue; this hiding place protects them from extreme temperatures and attack by other microorganisms. Some can also survive on various weed species. Spores spread among plants and from leaf to leaf by splashing water and on air currents—or by hitching rides on workers' hands or insects' feet.

Management

Leaf spot management tactics depend not only on the potential damage the disease can cause, but also on whether you are growing herbaceous perennials for pleasure or profit. Most leaf spot diseases do not kill plants and need not automatically prompt a control action in the garden. A level of damage that is tolerable in a home or public landscape will sometimes require more stringent management in nurseries and garden centers, however.

The most effective form of leaf spot control is genetic resistance. Selecting resistant species or varieties can minimize or even remove the need to use other tactics. Unfortunately, resistance is not available for many leaf spot diseases.

Since leaf spot fungi thrive on prolonged wet periods, strategies to keep the foliage dry will reduce disease risk. Watering at the soil line, such as by drip irrigation, rather than by overhead irrigation, is helpful. If you do use overhead irrigation, it is best to water in late morning to midday so that leaves can dry quickly. Spacing plants well apart, rather than crowding, will allow good air movement and speed up drying.

In the nursery, examine new plants as they arrive and remove any with obvious leaf spot symptoms. To disrupt the disease cycle in a field or landscape bed, remove as much diseased foliage as possible, both during the growing season and after foliage dies back in the fall. Destroy or dispose of this infested plant material; never use it as mulch in commercial plantings or the garden. By denying leaf spot fungi their hiding places in dead leaves, you can reduce the risk of future leaf spot outbreaks. Thoroughly compost diseased leaves and stems before returning them to the garden.

Fungicide sprays can be used to protect foliage in situations where the risk of substantial leaf spot damage is high. Fungicides will be most effective when applications are made preventively, before symptoms have appeared. Discontinue sprays in dry weather unless overhead irrigation is used frequently.

Alternaria leaf spot on farfugium.

Gray Mold

Botrytis flower blight on gladiolus.

INTRODUCTION

The gray mold fungus, *Botrytis cinerea,* is found almost everywhere plants are grown. It is fast growing, uses many different sources of nutrients, survives well in the greenhouse and outdoors, and attacks many different types of plants. The disease is also commonly called Botrytis blight.

Symptoms and signs

Botrytis at first appears as a white growth on the plant but soon darkens to gray. Smoky-gray, dusty spores are spread by the wind or splashing water. Infections can appear on wounded tissue such as large stubs left after taking cuttings, fading flowers, leaves onto which fading infected flowers have fallen, broken stems or injured leaves, seedlings grown under cool, moist conditions, and the base of cuttings, especially those taken from plants with heavy infestations of *Botrytis.*

Infected plant tissue usually turns light brown or has a somewhat wet appearance. If some of the tissue is placed on a wet paper towel in a plastic bag at room temperature, *Botrytis* will form its gray, dusty spores within 3-5 days.

A different species of *Botrytis* causes a unique set of symptoms on tulip and lily that are referred to as "fire." Stunted shoots, also know as "fireheads," emerge with blighted, twisted, tightly rolled leaves that have a ragged or withered appearance. The oval to round, yellow to grayish brown leaf and stem spots have dark, water-soaked borders. Flower buds can also become spotted, and badly blighted ones may fail to open.

Ecology

Any activity in or around infected stems and foliage will release a cloud of spores. Even turning on an automated trickle irrigation system can trigger spore release. The spores can remain dormant on plant surfaces as long as the life of

Tulip fire caused by *Botrytis cinerea.*

Botrytis leaf spot on lady's mantle.

the plant in some cases. *Botrytis* forms two types of resting structures on or in infected plant tissue: very dark brown or black, multi-celled structures called sclerotia, and single-celled, thick, dark-walled chlamydospores. It rarely undergoes sexual reproduction.

Botrytis thrives when it can absorb nutrients before invading a plant. These nutrients may leak from wounded plant parts, from dying tissue such as old flower petals, or from dead plant parts left over from the previous season. Pollen may also provide nutrients to *Botrytis.* Prolonged periods of high humidity, along with mild to warm temperatures, favor the growth of this fungus.

Management

Sanitation is the first important step. Remove dead or dying tissue from the plants and from the soil surface. This is particularly important outdoors so that new plants emerging in the spring do not immediately come into contact with *Botrytis.* Sanitation greatly reduces the amount of fungus in the immediate vicinity. But the fungus can produce 60,000 or more spores on a piece of plant tissue the size of your small fingernail, so additional steps are needed.

It is helpful to avoid injuring plants in any way. Do not leave large stubs of tissue on stock plants when taking cuttings.

Keeping the humidity low around the plants is crucial to suppressing *Botrytis.* Heating and ventilating greenhouses, particularly early in the day when moisture has condensed and before sunlight has warmed the air, and again in the evening before sunset, can suppress *Botrytis* quite effectively. Outdoor plantings should be planned to provide good air circulation patterns with adequate spacing between plants. Avoid placing highly susceptible varieties in locations where dew and precipitation remain on the plant for extended periods. Where possible, sprinkler irrigation should not be used; preferably, direct the water to the soil and keep the foliage dry.

Fungicides or biological control agents provide an added margin of protection when plants are particularly susceptible or the environment is especially disease-favorable. Keep in mind that some greenhouse and outdoor populations of *Botrytis* are resistant to certain fungicides—for example, benzimidazoles such as thiophanate methyl—so make sure to rotate fungicides with different modes of activity against the fungus. Resistance to dicarboximide fungicides such as iprodione and vinclozolin is less common but does occur. Biological control agents, like chemical fungicides, must be applied repeatedly in order to maintain adequate protection.

Powdery Mildew

Powdery mildew coating stems of peony.

INTRODUCTION

Powdery mildew occurs on many herbaceous perennials and annuals, as well as on woody ornamentals and trees. Although some powdery mildew fungi attack only one or two different species of plants, others attack many herbaceous perennials. All the powdery mildew fungi are obligate parasites, so they can grow and reproduce only on live plant tissue. In greenhouses, these fungi persist by spreading from a crop of diseased plants to a new crop that is susceptible to the same powdery mildew species. If the greenhouse is crop-free for several weeks, the fungus dies out and diseased plants must be brought into the greenhouse to establish disease again. Outdoors during the growing season, the fungus spreads long distances when the spores are carried by wind.

Phlox leaf showing the overwintering structures of powdery mildew.

Heavy powdery mildew infection on rudbeckia foliage.

Upper and lower surfaces of dahlia leaves may be coated by powdery mildew.

Symptoms and signs

White powdery fungus grows mainly on the upper surface of the lower leaves, but also on stems and flowers. Sometimes the patches coalesce to completely coat an infected leaf. In some cases leaves may be twisted and distorted, then wilt and die. On many plants, the older infected foliage falls off and younger foliage becomes infected later in the season. On some plants, a reddish purple discoloration can occur without visible sporulation or mycelial growth. At the end of the growing season, a close observer may see, on the surface of the white patches of powdery mildew, tiny yellowish globes that darken to dark brown or black. These are sexual spore cases, called chasmothecia or cleistothecia.

Ecology

Powdery mildew outbreaks are cued by extended periods of high relative humidity at night, which trigger formation of new spores. Low relative humidity during the daytime favors the spread of the spores in rising air currents. Temperatures of 70-80° F are ideal for rapid disease development.

The airborne spores land on leaves and germinate under humid conditions. Powdery mildew fungi grow primarily on the leaf surface but send fine threads (haustoria) between the cell wall and the cell membrane to obtain nutrients.

Management

In some species of herbaceous perennials, varieties differ widely in resistance to powdery mildew. It pays to notice which varieties are free of powdery mildew every year; planting these varieties in the future will be excellent insurance against the disease. When susceptible varieties are grown, they should be monitored on a regular basis in order to detect the first signs of mildew. In greenhouse and nursery production, it is best to group mildew-susceptible varieties so that scouting can be done efficiently and spraying, if necessary, can be confined to a small area. When signs of powdery mildew are first observed, apply a fungicide, horticultural oil, soap, or biological control agent to protect the plants. Powdery mildew is not usually lethal to plants, so in many cases the disease can be tolerated to avoid the inconvenience of treatment. In the perennial border, other plants may be grown in front of powdery mildew-prone species to hide the unsightly foliage late in the season. Since liquid water on leaves inhibits spore germination in most powdery mildews, syringing the leaves during low-humidity times of day can inhibit infection and suppress powdery mildew to some extent. Syringe only if other leaf diseases are not a problem, since other pathogens require liquid water to infect. Syringing may be the best approach at locations where fungicide use is not feasible or desirable and on crops for which there are no registered fungicides. Since some powdery mildews have developed resistance to certain systemic fungicides, do not rely solely on a single systemic fungicide, but rotate among fungicides with different modes of action in successive sprays and utilize contact action fungicides (such as bicarbonates or horticultural spray oil) as well.

Root Rots Caused by Fungi

Roots of gaillardia damaged by *Thielaviopsis* and *Rhizoctonia* spp.

INTRODUCTION

Root rots of perennials kill plants, but can also slow or stop plant growth. Affected plants are usually smaller, less vigorous, and produce fewer and/or smaller leaves, flowers, and fruit than healthy plants of equal age. Flowering may be delayed when the plant's roots are rotted. As a result, the quality of the crop or planting is very uneven. Root rots need to be prevented or managed at a very early stage of the disease if serious losses are to be avoided.

Management in established beds begins with making an accurate diagnosis of the cause of the root rot, since appropriate defensive action depends on knowing which pathogen is causing the problem. Most root rot fungi tend to be localized in an area of a nursery or plant bed—a "hot spot"—rather than widespread. They live in the soil and can spread several feet by soil splash, or many miles when infected plants or infested soil particles are moved from place to place by erosion or human activity. Root rots are a common problem in nursery container culture.

Symptoms

Fungal root rots cause a wide range of symptoms, and are often mistaken for other plant problems. A common symptom is slower growth in comparison to healthy plants. Root problems may also result in nutrient deficiencies that cause yellowing or reddening of leaves. Margins of leaves may turn brown, and older leaves may fall off. Roots appear dark brown or black, and may have a limp texture, unlike the white, crisp-textured appearance of healthy roots. When plants are pulled from the soil, the outer layer of cells sometimes strips off the roots, leaving only the central strand of water-conducting tissue.

In addition to fungal root rots, these symptoms can also be caused by too much or too little fertilizer, over- or under-watering, root exposure to chilling or freezing temperatures, phytotoxicity due to misapplication of pesticides used as soil drenches, and even the damage resulting when a plant has become pot-bound.

Before any action is taken, a reliable diagnosis must be made. If the damage is due to the activity of one or more fungi, applying appropriate fungicides can check the pathogens and allow the plant to grow. The fungus is usually not killed by fungicides, but its activity is slowed or stopped as long as the fungicide is in high enough concentration. Therefore, repeated applications of fungicides may be necessary.

Many fungi can cause root rots. Often, it is possible to identify which fungus is responsible either by observing the structure of the fungus in the roots using a microscope, or by placing infected roots on artificial media or baits (apple, carrot, or potato pieces) and allowing the fungus to grow out where it can be detected and then identified. This should be done at a plant diagnostic clinic. It is not uncommon for a plant to be infected with more than one root-rotting fungus at a time. The following are profiles of the most common genera of root-rotting fungi:

Rhizoctonia prefers warm, moderately moist soil conditions (70 to 90° F and 65% soil saturation). The fungus persists in soil for long periods of time, even if susceptible plants are not present. Plant tissues decay quickly, causing brown to reddish brown lesions or large spots to form at or just below the soil line. As these spots enlarge, they form sunken cankers that may eventually girdle the plant. Infected tissue may have a dry, shredded appearance. Soil particles often cling to the cankered areas of the plant when removed from the soil because of the coarse, brown mycelium. Leaves touching the soil can be infected. In high-humidity conditions, the fungus will quickly form webs to neighboring leaves and rot them. Under magnification, hyphal branches are at 90-degree angles to the parent hypha and there is a cross-wall and a constriction of the cell at the base of each branch. *Rhizoctonia* does not form spores. It moves about the garden in soil that is contaminated with fungal mycelia or sclerotia (tiny knots of fungus that provide long-term survival in soil).

Thielaviopsis, which causes black root rot, is favored by wet soils, soil pH above 5.6, and temperatures between 55 and 65° F. Distinctive dark brown spores in the infected roots are readily observed with a microscope. These dark spores resist high and low temperatures and moisture extremes, allowing *Thielaviopsis* to persist in soil for long periods of time. The individual cells of these dark spores appear to snap apart. Light-colored spores are also formed in a long tapered cell and extruded in chains.

Fusarium species form distinctive canoe-shaped spores with several cross-walls. Masses of these spores are sometimes visible as white, dusty growth on the lower stems and upper root system of infected plants. *Fusarium* species can also attack bulbs, tubers, and corms. Infected roots may initially have a reddish color but soon darken. Some

Sclerotia of *Sclerotinia sclerotiorum* inside a stem.

Sclerotium rolfsii symptoms and sclerotia on alstroemeria.

plants develop secondary roots above the infected area but these too become diseased. *Fusarium* attack is favored when plants are weakened by over-fertilization, drought, mechanical injuries, or other stresses.

Sclerotium species attack roots and stems near the soil line during warm, wet periods. In North America, disease caused by *Sclerotium rolfsi* is sometimes called Southern blight. The pathogens on herbaceous perennials, mainly *Sclerotium rolfsii* and *Sclerotium delphinii,* are transported long distances on infected plants, and can thereby be introduced into greenhouses located in colder climates. Succulent plants are killed quickly and a white to cream-colored, fan-like mat of fungal growth forms on the lower stem, leaves, and soil close to the plant.Clusters of round ball-like masses of fungus (sclerotia) form on the surface of this material. These balls of fungus turn light to dark brown and persist for long periods on the infected tissue and in soil.

Sclerotinia causes a rot of stems and roots near the soil line that is commonly called white mold, crown rot, or cottony rot because of the fluffy, white fungal growth that forms at or near the soil line. Knots of fungal growth (sclerotia) within the masses of fungus seen on the plant are soft and white at first but become hard and black with time. Usually the first symptoms noticed are yellowing and collapse of the entire plant.

Phymatotrichopsis (formerly called *Phymatotrichum*) attacks a very wide range of plants in hot, dry climates, particularly in soils with high pH, high clay content, and poor aeration. A distinctive sign of this fungus is formation of thick strands of coarse, brown fungal growth along the infected roots. Long-lasting resting structures (sclerotia) allow *Phymatotrichopsis* to persist in the soil for long periods of time. Although it is killed by freezing temperatures, it can survive deep in soil below the frost line.

Pythium and ***Phytophthora***, two important fungus-like genera, are discussed elsewhere in this book. They commonly cause root rots as well.

Management

The most effective strategy is to eliminate root-rotting fungi from the planting area or the potting soil before introducing plants. In greenhouse production, heat treatment (usually steam pasteurization) is the best method of freeing potting soil of root-rotting fungi. Although no single fungicide will kill or inhibit the growth of all fungi, some are effective against many genera of fungi, and therefore a different fungicide is not needed for each and every fungus.

When a root rot problem is suspected, work with a diagnostic clinic to find out which fungus or fungal group is causing the problem. This is especially important since symptoms of many root rot diseases and physical stresses can look alike. Get a reliable recommendation of which fungicide is effective in managing that particular fungus or fungal group. The fungicide's label usually indicates the fungi against which it is effective.

Rusts and Smuts

INTRODUCTION

The rusts and smuts belong to the Basidiomycota, the group of fungi that includes the mushrooms.

Rust fungi attack a large number of herbaceous perennials. The common name "rust" comes from the reddish to orange, rusty-colored pustules, or blister-like swellings, that rust fungi produce on leaves, scapes, or stems of host plants. Rust fungi are obligate parasites, which means that they need living plants to survive, and therefore seldom kill plants. Nevertheless, rusts harm ornamental plants' appearance by reducing flower production, and detracting from overall plant health and vigor. Severe infections can result in premature leaf drop.

Smut diseases are far less common than rusts on herbaceous perennials, and are seldom major concerns. The smut fungi can infect leaves, stems, and flower parts depending on the smut species, host plant species, and timing of infection. The smuts are named for their sooty black spore masses, which commonly replace infected plant parts.

RUSTS

Symptoms

Rust symptoms are different on the primary and alternate plant hosts. On leaves, scapes, or stems of most herbaceous perennials, rust fungi produce blister-like swellings, called uredinial pustules, which may be reddish, orange, yellow, white, or dark brown, depending on the species of rust. On alternate plant hosts, these same fungi can produce aecial pustules that are typically larger than uredinial pustules and are usually yellowish in color.

Ecology

While most parasitic fungi have only one or two spore stages and a single plant host, rusts lead more complex lives. Some rusts have as many as *five* spore stages and *two* plant hosts.

For some rusts, two types of host plants are needed to complete the disease cycle. These dual-host rusts produce specialized spores called urediniospores on one of the host plant species, and another type of specialized spores called aeciospores on the other plant host. Rust diseases with this lifestyle are said to have "alternate hosts".

For most rust diseases of herbaceous perennials, the perennial plant is the host on which the urediniospores form. Blister-like pustules called uredinia break open on the plant surface to expose the rusty-colored urediniospores. This stage is the most damaging to the plant: since urediniospores can cause new cycles of infection in

Brown rust on chrysanthemum.

Rust pustules on Helianthus.

Rust pustules and urediniospores on daylily.

Teliospores forming in daylily rust pustules.

the same season, rust epidemics can can develop rapidly under favorable weather conditions. During late summer, rust fungi switch to a survival strategy, resulting in production of black, thick-walled teliospores that remain dormant during the winter. The following spring, teliospores germinate to produce yet another type of spores called basidiospores.

The basidiospores infect another species of plant, which is often not closely related to the one on which the urediniospores, teliospores, and basidiospores were produced. Shortly after infection by the basidiospores, small vase-shaped structures, called spermogonia, are formed. They protrude through the upper epidermis of the leaf of the infected plant. Some weeks later, a cuplike structure called an aecium bursts through the lower epidermis. The aeciospores produced in the aecia move by air currents and splashing water to infect the original host plant, and the dual-host cycle starts over again.

Rust spores move easily on wind currents and in splashing water, but long-distance movement happens mainly in shipments of infected plants. Many rusts survive the winter in warmer climates, where host plants stay alive year-round, and spread throughout the U.S. each spring and summer on shipments of infected plants. Shippers often transport the rusts accidentally, because microscopic rust spores may be clinging to healthy-looking plants.

White smut symptoms on gaillardia.

Management

Integrated management practices, including scouting, proper sanitation, use of resistant varieties (when available), and preventive fungicide applications, are used to manage rust outbreaks and minimize losses. Several chemical classes of fungicides are effective against rusts on ornamental crops.

The most important recommendation is to avoid introducing rust diseases into your garden or nursery. First, buy only from reputable sources that offer certified, inspected, or pathogen-free plants. Carefully inspect all plants for any evidence of disease before buying and planting. If possible, separate newly purchased plants from the rest of a production house, field, or garden until you are confident no rust will develop. This may be particularly important for daylilies, which have been troubled by a rust in the United States since the beginning of this century.

If you have infected plants, one option to slow or eliminate the spread of rust is removing and destroying infected plant material promptly, so the pathogen cannot be transmitted to other healthy plants in the surrounding area.

SMUTS

Symptoms

Symptoms vary with the type of smut that is infecting the plant. Plants with leaf or flower smut may be stunted, yellow, or malformed. To accurately confirm a diagnosis of smut, look for powdery black masses of teliospores in infected tissues. Some perennials are infected by white smut; the white color is from the sporidia produced from the teliospores.

Ecology

Smuts produce two types of spores, called teliospores and sporidia. Teliospores survive periods of harsh environmental conditions on contaminated seed, in plant debris or in the soil; they eventually germinate to produce sporidia, which can bud to produce more sporidia. Infection on seedlings originates from spores carried on seeds or in the soil. Short-distance movement of spores is by wind and long-distance movement is through infected seed and plant parts.

Management

The best management option is to remove and destroy infected plant material, which slows or eliminates the spread of smut. A number of fungicides are labeled for control of smut, but must be applied preventively (before symptoms appear) to be effective.

Vascular Wilts Caused by Fungi

INTRODUCTION

Vascular wilt diseases are caused by fungi or bacteria that clog the fluid-conducting tissues of plants, resulting in stunting, discoloration, wilting, or death. The most important vascular wilt diseases of herbaceous perennials are caused by fungi (*Fusarium* and *Verticillium*). Some bacteria, such as *Ralstonia solanacearum,* also cause wilt symptoms by infecting the vascular system (see section on Bacterial Diseases).

FUSARIUM WILTS

These diseases are caused by specialized strains, also known as "formae speciales," of the fungus *Fusarium oxysporum.* Different *F. oxysporum* strains cause wilt diseases on astilbe, dianthus, dahlia, lisianthus, hibiscus, *Ranunculus,* chrysanthemum, bleeding heart, purple coneflower, *Sedum, Nepeta,* and many other herbaceous perennials. The host ranges of many *F. oxysporum* formae specialis are restricted to a single species of plant.

Symptoms

Affected plants are chlorotic and become wilted. Plants may be stunted, and older leaves, especially, may show scorch symptoms. Highly susceptible varieties may die, while others may show more moderate symptoms. Severe symptoms often develop after a period of hot weather. A cross or longitudinal section cut near the base of the stem often reveals a brown, olive green, or reddish brown discoloration just beneath the surface.

Ecology

Fusarium oxysporum, a soil inhabitant, typically invades plant roots. The fungus then moves into the xylem vessels, where it spreads in the transpiration stream throughout the plant. Eventually the fungus grows out through the stem layers to the plant surface, where new spores spread easily in greenhouses or outdoor plantings by splashing water, handling, or propagation from infected plants.

Fusarium wilt is most destructive in warm climates and in warm, sandy soils. In the U.S., the disease is most severe on outdoor plantings in southern and central states. High greenhouse temperatures can also trigger outbreaks. Between crops, *F. oxysporum* survives in infested crop debris as mycelium or chlamydospores. The fungus spreads short distances in flowing water or on contaminated tools or equipment, but long-distance transport is mainly on infested nursery stock or in the soil attached to them. Once a field becomes infested with *F. oxysporum,* it is likely to remain that way indefinitely.

Fusarium wilt on chrysanthemum.

Advanced Fusarium wilt symptoms on chrysanthemum.

Nutritional factors can influence development of Fusarium wilt. High levels of ammonium nitrogen encourage the disease, especially when potassium levels are insufficient. On the other hand, nitrate nitrogen has some suppressive effect on the disease. Low calcium levels may also increase disease severity.

Management

The widespread adoption of soilless media has greatly reduced outbreaks of Fusarium wilt fungi in greenhouses. Outdoors, liming soils and using nitrate nitrogen have suppressed *F. oxysporum* effectively on chrysanthemum, aster, and gladiolus. Crop rotation to non-host species can control the disease, but rotating back to host crops is likely to be risky because of the pathogen's ability to survive long periods in the absence of its hosts. Although fungicide drenches may inhibit the pathogen to some extent, fungicides are generally ineffective against Fusarium wilt.

VERTICILLIUM WILTS

Two species of soil-dwelling fungi, *Verticillium dahliae* and *Verticillium albo-atrum*, attack more than 200 species of plants, including many herbaceous perennials. *V. dahliae* prefers slightly higher temperatures (77-82° F) than *V. albo-atrum* (68-77° F), and is somewhat more common in warmer regions. Unlike Fusarium wilt, the Verticillium wilt fungi are not highly host-specific on herbaceous perennials.

Symptoms

Stunting, slow growth, leaf scorch, chlorotic foliage, and wilting—all typical symptoms of Verticillium wilt—result from the stress created as the fungus blocks upward movement of water. Symptoms usually begin on lower leaves and progress upward. A characteristic of Verticillium wilt is that wilting may initially develop on only one side of the plant. Infected plants may die if they are exposed to water stress. A light brown to black discoloration of the vascular system is sometimes present, but it is not a dependable symptom for diagnosis.

Ecology

Both *Verticillium* pathogens are soil natives, and can survive for many years in soil in the absence of a host plant. *V. dahliae,* which forms survival structures called microsclerotia, can survive without a host for as long as 15 years. Both species can also survive the winter in perennial hosts, in propagative organs such as bulbs, or in plant debris. Verticillium fungi penetrate young roots of host plants directly or through wounds. The fungi spread on contaminated seed, on vegetative cuttings and tubers, and by wind, splashing or flowing water, or in soil particles. Solanaceous crops like tomato, potato, and eggplant, as well as alfalfa, are highly susceptible to Verticillium wilt, so susceptible herbaceous perennial crops following these crops will be at particularly high risk of the disease. Nevertheless, these pathogens are native to many soils, so avoiding land used for these crops does not eliminate the risk of outbreaks.

Although the risk of Verticillium wilt in greenhouse production has ebbed with the predominance of soilless media, the disease can still be introduced into greenhouses, either when outdoor-grown plants are potted up and moved indoors or in shipments of symptomless but contaminated cuttings. Limited evidence suggests that certain fungus gnats can spread *V. albo-atrum* in greenhouses by introducing the fungus to their feeding wounds on roots, stems, or leaves. During composting of plant materials infested with the Verticillium wilt pathogens, achieving a temperature within the pile of at least 150° F is necessary to kill these fungi.

Management

Planting culture-indexed stock that is certified to be free of fungal pathogens is a valuable first step in reducing the risk of Verticillium and Fusarium wilt. Remove and destroy plants that are showing symptoms. If using field soil in greenhouse media, pasteurize the soil with steam before adding it to the mix. Fumigation of field soil can reduce populations of *V. dahliae* and *V. albo-atrum*, but may not be cost effective in many cases. Species and varieties vary considerably in resistance to Verticillium wilt; resistant types should be planted in fields with a disease or cropping history that indicates high risk of the disease. However, eventual rotation back to susceptible hosts is unlikely to work well due to the prolonged survival of the pathogens in soil. Fungicides are generally ineffective against Verticillium wilt. Liming soil to achieve a high pH (above pH 6.0) and using fertilizer that includes only the nitrate form of nitrogen have been helpful during chrysanthemum production.

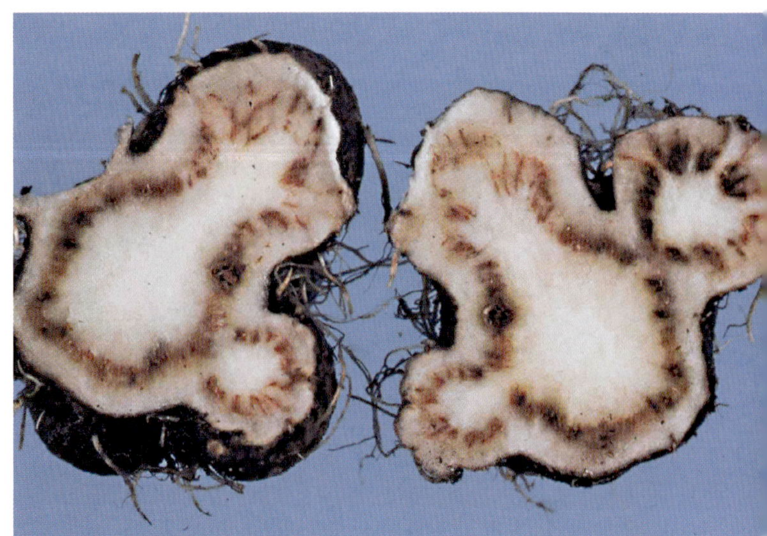

Discoloration of vascular ring on liatris by Verticillium wilt.

Downy Mildews

Downy mildew on rudbeckia.

INTRODUCTION

Downy mildews are caused by a group of highly specialized obligate parasites. Obligate means that they can grow and reproduce only in a living host plant, but downy mildews also produce various types of resting structures that allow them to survive when the host is dead. Despite their similar-sounding common names, downy mildews are distinctly different from the powdery mildews. Unlike powdery mildews, downy mildew organisms are not true fungi. Even so, they look and act similar to fungi in many ways. The chemicals used to control downy mildews are similar to those used against their relatives, *Pythium* and *Phytophthora*, but different from most of those used against true fungi.

Symptoms

Downy mildew colonies often appear first as small tufts of white to purplish, fluffy growth on the underside of the leaf; small yellow spots develop on the upper sides of the leaf. As the host tissue dies, the fluffy growth darkens to gray. Infected leaves and branches may be distorted and die. Fast-growing downy mildew colonies can be confused with gray mold (*Botrytis*), and slower-growing colonies with powdery mildew. Microscopically, downy mildew is very easy to tell apart from powdery mildew and *Botrytis*. Like bacterial and foliar nematode lesions, spots caused by downy mildews are often angular-shaped, but unlike those caused by bacteria, downy mildew spots do not have a water-soaked appearance in early stages of disease. Under a microscope, angular spots due to foliar nematodes may be distinguished by observing the nematodes swimming out from the infected tissue in a small dish of water.

Ecology

Most downy mildews thrive when cool temperatures (58-72° F) are accompanied by high humidity (above 95%). Some downy mildew species survive the winter in cold climates in infested plant debris, soil, or weeds; others must be dispersed northward from southern regions of North America each spring, often in weather systems.

Species of downy mildews reproduce sexually by structures called oogonia and antheridia, and asexually by sporangia. The sporangia germinate directly by forming a germ tube. In some genera (*Sclerospora* and *Plasmopara*), sporangia can also germinate indirectly by releasing specialized swimming spores called zoospores. *Peronospora*, *Pseudoperonospora*, and *Bremia* seldom form zoospores.

Management

Debris from perennial plants should be removed and buried, burned, or placed in a closed container. Plants should be spaced widely enough to ensure good air circulation and rapid dryoff. Overhead irrigation should not be used when the weather is generally cool. Fungicides similar to those used for *Pythium* and *Phytophthora* are available to control downy mildews.

The upper surfaces of leaves infected with downy mildew often show yellow, purple or brown patches of discoloration, as shown here on rudbeckia.

Downy mildew on geum.

Root and Stem Rots Caused by Fungus-Like Organisms

(*Pythium* and *Phytophthora*)

Pythium root rot on delphinium.

INTRODUCTION

The fungus-like pathogens *Pythium* and *Phytophthora* are actually more closely related to algae than fungi. Like true fungi, they lack chlorophyll and therefore must absorb nutrients from other plants. Because they are so different from true fungi, most of the chemicals used to control them do not control true fungi.

PYTHIUM

Pythium species that cause root and stem rot in many different perennial plants include *Pythium irregulare*, *Pythium aphanidermatum*, and *Pythium ultimum*. *P. ultimum* and *P. irregulare* are often found in field soil, pond and stream sediments, and dead roots of previous crops. *Pythium irregulare* and a closely related species, *Pythium cryptoirregulare*, have been isolated from many herbaceous ornamentals.

Symptoms and signs

- Plants are stunted.
- Root tips are brown and dead.
- Plants wilt at mid-day but may recover at night.
- Leaves yellow and die.
- Brown tissue on the outer portion of the root pulls off easily, leaving a thin core of vascular tissue exposed.
- The cells of roots contain round, microscopic, thick-walled spores.

Ecology

Pythium is widespread in field soil. In gardens, it will be a problem primarily in areas with poor drainage or excessive irrigation. In production, *Pythium* is easily introduced into soilless growing media and moved from one area to another on dirty tools, pots, or flats. Fungus gnat and shore fly activity can also move *Pythium* short distances. When introduced into a soil mix that has been heat-treated for too long or at too high a temperature, *Pythium* can cause severe root rot because of the elimination of competing microorganisms. *P. aphanidermatum*, *P. cryptoirregulare*, and *P. irregulare* pose a threat to crops grown in ebb and flow systems because they form zoospores that can swim in water. This is most likely to occur if irrigation water is captured and recycled. In greenhouse-grown perennials, *Pythium* problems can be very severe if irrigation times are prolonged (45 min. or longer) or if pots are allowed to sit in puddles of water on the bench or floor. If *Pythium* infests a cutting bed or if contaminated water is used in propagation, large losses may occur. *Pythium splendens* is a pathogen mainly on tropical plants. *Pythium ultimum* is primarily associated with soil and sand, so as growers switched to using soilless mixes for container production, this species became less important. *P. ultimum* does not form the swimming spore stage, but can be a problem in ebb and flow systems if the reservoir becomes fouled with potting mix and plant debris particles. Almost all plants are susceptible to Pythium root rot. Root tips, very important in taking up nutrients and water, are attacked and killed first. *Pythium* also can rot the base of cuttings. *Pythium* can be moved long distances on infected plant material shipped from place to place. Under some circumstances, plants can recover from *Pythium* infections.

Management

Pythium root rot is difficult to control once it has become established, so every effort should be made to prevent the disease before it begins. In container production, *Pythium* can be eliminated by using heat-pasteurized potting mix. Heat the entire pile to 180° F and hold it at that temperature for 30 minutes to kill all plant pathogens in the soil while leaving many beneficial fungi and bacteria alive. Long treatment times (an hour or more) and higher temperatures can kill beneficial organisms, and should be avoided. Once the potting mix has been treated, it should be covered and stored in a way that reduces the possibility that it will be contaminated through the introduction of non-treated soil. Commercially available soilless potting mixes are generally free of *Pythium*. These too should be handled and stored in a manner that prevents contamination by untreated soil or plant debris.

If pond or stream water is used for irrigation, the intake pipe should be located well above the bottom so that sediment is not drawn in. If the water supply is suspected of being a source of *Pythium*, it may be necessary to treat the water before use. Several different techniques can eliminate *Pythium*; select one after consulting with a water treatment expert. Slow sand filtration has been shown to be an effective, simple, and inexpensive method for removing *Pythium* from water. Heat, ultraviolet light, ozone treatment, and chlorination can be effective but require some training to be used properly.

In greenhouses using recirculating water systems, the ebb and flow system reservoirs should be covered to prevent contaminated debris from entering the system. Pass return water over a coarse screen to remove potting soil and plant debris in order to help keep *Pythium* out of the reservoir. Disinfect all bench surfaces, potting benches, tools, and equipment that will contact the potting mix. Periodically, thoroughly clean and disinfest ebb and flow reservoirs, benches, and flood and drain floors; pay particular attention to cracks and corners where infected plant debris can become lodged.

In a greenhouse or outdoor operation with a history of Pythium root rot, apply a fungicide or a biological control agent as early in the cropping cycle as possible. Biological agents should be applied before, during or immediately after transplanting. It is generally recommended that chemical pesticides not be applied to the soil or potting mix for 10 days before and after applying the biological control agent. Biological control agents and fungicides may need to be applied more than once in order to maintain adequate protection.

Some populations of *Pythium* have resistance to the fungicides metalaxyl, mefenoxam and/or propamocarb. Chemical control programs may include these, but should use them within a rotation that includes other materials such as etridiazole for which resistance is not known.

PHYTOPHTHORA

Several *Phytophthora* species cause root rot of perennials including *Phytophthora nicotianae*, *Phytophthora cryptogea*, and *Phytophthora cactorum*. While they are often referred to as being 'soil-borne' (residing in soil even in the absence of plants), most species of *Phytophthora* are actually 'plant-borne;' in other words, they stay associated with infected plants or in plant debris from infected plants and do not survive extended periods (years) free in the soil. *Phytophthora* can form a swimming spore stage (zoospores) that can move in water and cause severe losses when the infested irrigation water is captured and recycled. In greenhouse-grown perennials, *Phytophthora* problems can be very severe if irrigation times are prolonged (45 min. or longer) or if pots sit in puddles of water because the bench or floor drains poorly.

Symptoms

Plants with mild root rot have smaller than normal foliage, dead feeder roots, and/or dark streaks up the stem. Severe root rot can produce stunted and/or wilted plants, smaller than normal leaves, greatly reduced root systems with rotted, reddish-brown discoloration, lack of new shoot development, or death. Plants with stem rot may show wilting, a darkened and water-soaked area near the soil line, collapse of the stem, and/or secondary infection by bacteria.

Symptoms of Phytophthora crown rot on agapanthus.

Phytophthora crown rot on brunnera.

Ecology

Infection can occur under mild to warm temperatures (59 to 82° F), depending upon the species of *Phytophthora* involved. When the soil is wet, sporangia (thin-walled bags of zoospores) form in 4-8 hours and zoospores can be released 10 to 60 minutes later. Prolonged periods of water-logged soil favor the spread of *Phytophthora*. *Phytophthora* survives the winter in infected roots and other plant parts. It can be splash-dispersed during heavy rains or overhead irrigation or carried in run-off water from plant to plant in the field or in container production.

Management

Disease prevention is the main goal, since no chemicals cure *Phytophthora*-infected plants. An important starting point is to avoid buying plants that may harbor *Phytophthora* infections; it makes good sense to inspect all incoming stock for symptoms, and discard any suspicious-looking plants. Several chemicals are available to protect plants but there are no biological control agents that provide adequate control of *Phytophthora*.

In production fields, it makes sense to plant only in well-drained locations. If the area previously harbored *Phytophthora*, fumigation before planting will greatly reduce the amount of the pathogens in the soil. Fumigate when soil temperatures are 10° C (50° F) or warmer at 15-cm (6-in) depth and when soil moisture levels are adequate for seed germination. Allow adequate aeration time after fumigating with chemicals. Fumigation should be done only by commercial nursery producers and landscapers, since the fumigants used are highly toxic and dangerous to humans as well as *Phytophthora*. Gardeners and nurserymen alike should avoid overhead watering, especially in late afternoon. In fields with a history of *Phytophthora* problems, treat transplants immediately with an appropriate chemical, repeat the application at the recommended interval, and avoid the use of recirculating irrigation systems.

In container production, use a well-drained, pathogen-free potting mix. For example, composted hardwood bark not only drains well, but also inhibits *Phytophthora* by means of natural biological control. Use clean containers. Place them on an area that was graded to insure good surface drainage, or on a 4–inch-thick bed of gravel or other well drained material. Black plastic under this bed will prevent weed growth, but avoid using impermeable black plastic directly beneath containers, as this will cause puddling. It is important to group different types of plants by water requirement so that plants are not over- or under-watered. In nurseries with a history of *Phytophthora* problems, treat transplants immediately with an appropriate chemical and repeat the application at the recommended interval.

In the landscape, plant only in well-drained areas, or grade or tile the site to insure good drainage. Avoid planting in locations where plants were infected previously by *Phytophthora*. Do not plant too deeply; the soil line should not be more than 1 inch over the upper roots.

Bacterial Diseases

Xanthomonas leaf spot on peony.

INTRODUCTION

Bacteria are microscopic, single-celled organisms that have a cell wall. Their genetic material, a circular strand of DNA, is not surrounded by a nuclear membrane. Therefore, bacteria do not have a true nucleus as do plants, animals, and fungi. While most bacteria in the environment are beneficial, several are able to cause leaf spots, stem rots, root rots, galls, wilts, blights, and cankers.

Symptoms

Symptoms of bacterial diseases vary widely. For example:

Agrobacterium tumefaciens causes a disease called crown gall on many herbaceous perennials and other plants. Galls, which are tumor-like swellings of plant tissue, range from 1/4 inch to several inches in diameter; they typically form at the soil line but also can form on branches or roots. Galls are initially white, spherical, and soft but darken with age as outer cells die. The bacteria survive in the soil for many years.

Dickeya chrysanthemi and *Pectobacterium carotovorum* survive in plant debris that is not completely decomposed, in infected plants, on other greenhouse plants without causing disease, and in soil or potting media. Both species infect a wide range of plants in the greenhouse. They can cause either leaf spots or a mushy, brown, smelly, soft rot of all plant tissues.

Pseudomonas cichorii can cause leaf spots and blights on chrysanthemum, geranium, impatiens, and many other ornamental and vegetable plants. The spots are generally water-soaked (wet-looking) and dark brown to black. Depending upon the plant infected and cultural conditions, the leaf spots may have a yellow halo.

Xanthomonas campestris pv. *pelargonii* causes bacterial blight, wilt, or leaf spot of pelargonium and hardy geranium. Other species and pathovars (abbreviated "pv.") of *Xanthomonas* attack a wide variety of other plants. Each pathovar of *Xanthomonas* or other bacterial pathogens typically attacks only one or a few species of plants.

Rhodococcus fascians causes fasciation, the formation of multiple small branches and stems that are flattened or distorted. Fasciation usually develops at or slightly below the soil line near the base of infected plants. Often the symptom caused by *R. fascians* is best described as a "leafy gall". The bacterium is carried on infected cuttings and may enter the propagation medium.

Ralstonia solanacearum, which causes vascular wilting of many herbaceous ornamentals, survives for long periods in soil. Symptoms include leaf wilting, discoloration of the vascular tissue, leaf yellowing and death of the plant.

Ecology

Most plant pathogenic bacteria survive in infected plants and in debris from infected plants. A few can also survive in soil. Most bacteria can invade plants only through wounds or natural openings (such as stomates or hydathodes) and require warm, moist conditions to cause disease.

Bacteria reproduce very rapidly. They are splashed easily from the soil to the leaves and from leaf to leaf by overhead irrigation or rainfall. They are also easily transferred when workers handle contaminated soil or plant debris, then handle live plants.

Soft rot caused by *Pectobacterium* sp. on hosta.

Management

The most effective way to avoid bacterial diseases is to buy only plants that are certified to be free of the pathogens by a process called culture indexing. In this procedure, pieces of plant tissue are incubated in a nutrient broth that will encourage the growth of plant pathogenic bacteria. If the test is repeated two to three times and no pathogenic bacteria are detected, the plant is said to have been culture indexed. Unfortunately, culture indexing is not yet widely available for herbaceous perennials other than chrysanthemums.

Sanitation practices are very helpful in controlling bacterial diseases. These include destroying infected plants, cleaning and disinfesting tools, benches, flats, and pots, and treating potting soil to to kill all pathogens. Soil in which infected plants were grown or rooted should be

Shoot proliferation on dahlia, caused by *Rhodococcus fascians*.

discarded or thoroughly steam-treated. Plant handling procedures and debris/soil handling operations should be done completely separately.

A key cultural practice is keeping the foliage dry when you irrigate. Overhead irrigation should be avoided in crops that are particularly susceptible to bacterial diseases. When there is no alternative to overhead watering, it should be done early in the day so that foliage dries as quickly as possible. To insure good air circulation within the crop canopy, outdoor plants should be spaced widely enough to avoid a dense, closed canopy of foliage. In greenhouses, horizontal air flow, with rows of plants oriented parallel to the air movement, can greatly reduce relative humidity. Trickle irrigation and capillary mat watering can avoid creating favorable conditions for bacterial spread and infection. Some bacteria have been shown to spread in ebb and flow systems for greenhouse irrigation. Steps should be taken to filter crop debris out of the water and treat the water with chlorine, chlorine dioxide, hydrogen peroxide formulations, ultraviolet irradiation or heat to eliminate hitchhiking bacteria.

Once bacterial disease appears on plants, chemical control is not effective and plants should be discarded.

Stem rot of chrysanthemum caused by *Dickeya chrysanthemi*.

Phytoplasmas

Aster yellows virescence on purple coneflower.

INTRODUCTION

Phytoplasmas are single-celled microorganisms that live as parasites in the phloem of plants and cannot be cultured routinely on agar media in the laboratory. Previously phytoplasmas were known as "mycoplasma-like organisms" or "MLOs". They cause a number of diseases on herbaceous and woody plants, inducing symptoms that were originally thought to be due to viruses. Later the tiny, irregularly-shaped bodies of the phytoplasmas were visualized in the phloem with the aid of the electron microscope. Phytoplasmas are distinguished from bacterial cells by the absence of a cell wall, which causes them to be variable in shape. The parasitic plant dodder *(Cuscuta)* can be used to transmit phytoplasmas from one plant to another for study purposes. Some of the best known phytoplasma diseases are ash yellows, elm yellows, and lethal yellowing of coconut palms, all of which cause serious decline of their hosts. On herbaceous perennials, the best known phytoplasma disease is one with a broad host range, affecting plants in over 40 families: aster yellows, caused by *Candidatus* Phytoplasma asteris.

Symptoms

Aster yellows is characterized by a range of unusually striking disease symptoms: plants are stunted, foliage is yellowed, and flower parts develop many curious abnormalities. One common symptom is virescence, which refers to the greening of normally colorful flower petals. Additionally, flower parts may revert to leaf forms (phyllody), creating a very peculiar-looking floral display. Among perennials, these symptoms are seen especially often in purple coneflowers (*Echinacea*). Upright leaves and witches'-broom are seen in some cases of aster yellows.

Aster yellows on coreopsis.

Ecology

The aster yellows pathogen is vectored in nature and in horticultural settings by the aster leafhopper, *Macrosteles quadrilineatus*, and certain other leafhoppers. The disease may be carried over long distances by the aster leafhopper, and annually appears in gardens in the northern United States as this leafhopper migrates up from the South. The aster yellows phytoplasma may be harbored locally in perennial and annual weeds such as Queen Anne's lace, dandelion, and horseweed, and can be carried into ornamental plantings each season once a leafhopper vector is available. It takes several weeks of incubation before a leafhopper can transmit the phytoplasma after acquiring it, and for most plant hosts it takes several weeks before symptoms are apparent after a leafhopper vector feeds on it. *Candidatus* Phytoplasma asteris has many hosts, including weeds as well as ornamentals and vegetables. Aster yellows appears only sporadically in ornamental plantings in most regions. It is not carried through seed. Susceptible plants include *Aster, Anemone, Centaurea, Chrysanthemum, Coreopsis, Delphinium Echinacea, Gaillardia, Limonium, Phlox, Scabiosa,* and *Veronica*.

Management

Managing the aster yellows phytoplasma in herbaceous perennials is primarily a matter of eradicating it promptly from nurseries or plantings once it is detected. There is no treatment for an infected plant. Spraying for vector control is not often effective, as the leafhoppers may be very abundant, very mobile, and transmit the pathogen before they are killed. Growers of the highly susceptible China aster (*Callistephus*) have in the past employed screening over cut-flower plantings to exclude the leafhopper. Yellow sticky cards may be used to monitor for the aster leafhopper's arrival in the northern areas of the United States. Weed control in the vicinity of the nursery or garden is important for eliminating reservoir hosts of the disease. Failure to detect this disease promptly may result in increasing problems over time. Rogue out the affected individuals as soon as possible. Select plants to grow that are not hosts of aster yellows if this disease becomes a significant problem in your area.

Virus Diseases

Symptoms of *Tobacco rattle virus* on anemone.

INTRODUCTION

Viruses are tiny submicroscopic particles that can cause disease. They are composed of nucleic acid (genetic material) surrounded by a protein coat. The nucleic acid in most plant-infecting viruses is ribonucleic acid (RNA); only a few contain DNA (deoxyribonucleic acid). Viruses are entirely dependent upon the host plant for their reproduction. Once inside the plant cell, the protein coat peels off and the nucleic acid portion organizes the cell to produce more copies of the virus. Energy and resources that normally go into producing more healthy plant tissue are diverted to producing virus instead, disrupting the normal activity of the cell. Viruses can multiply only inside a living cell.

The name of a virus often refers to the plant in which it was first found and the type of symptoms it causes. The name is not an indication of what other plants might be susceptible to it. Some viruses can infect hundreds of genera of plants while others only infect a few species. It is also not uncommon for an individual plant to be infected with more than one virus.

Symptoms

Symptoms vary with the virus or viruses involved, the species of plant infected, and the environment. Certain environmental conditions favor symptom development, while other conditions suppress symptoms. In fact, many virus-infected plants show no obvious symptoms at all. Symptoms associated with virus infections include the following: stunting; mosaic pattern of light and dark green (or yellow and green) on the leaves; distortion of leaves or growing points; yellow streaking of leaves (especially in monocots); yellow spotting on leaves;

Cucumber mosaic virus (CMV) on lobelia.

Hosta virus X on hosta.

ringspots or line patterns on leaves; cup-shaped leaves; uniform yellowing, bronzing, or reddening of flower or foliage color; distinct yellowing of veins only; and crinkling or curling of leaves or leaf margins. Production of non-uniform color in plants known for uniform color, referred to as breaking, is sometimes caused by virus infection; this symptom is often seen in flower petals.

Symptoms that mimic virus symptoms can also be caused by high temperatures, insect feeding, mineral shortages or excesses, and phytotoxicity from growth regulators or pesticides. Because of these look-alike problems, it is usually not possible to determine that a plant is infected with a virus based on symptoms alone.

Ecology

Most viruses, including *Cucumber mosaic virus* (CMV) and *Impatiens necrotic spot virus* (INSV), die quickly if they are outside a cell or if the cell dies. An exception is *Tobacco mosaic virus* (TMV), whose particles survive for years on tools, plant ties, or other surfaces where sap from an infected plant dries.

A virus cannot spread from plant to plant without help. Viruses can spread in one or more of the following ways, depending upon the specific virus involved: mechanically in plant sap on workers' hands or on tools; by aphid, thrips, whitefly, leafhopper, mite or nematode feeding; through dodder (*Cuscuta*) parasitizing the plants; through grafting from an infected plant; by pollen; through infection by certain soil fungi, or through shared water in recirculating irrigation systems.

Vegetative propagation spreads all virus diseases. Cuttings taken from an infected plant usually are infected even if no symptoms are showing. Rhizomes, corms, bulblets, stems or any other parts removed from an infected plant to start a new plant are likely to carry the virus, since it invades all parts of the plant except the few cells at the tips of the growing points. If these few cells are carefully removed (during the process called meristem tip culturing) in order to propagate a new plant, it is sometimes possible to obtain a new plant free of virus infection.

Management

To manage a virus disease effectively, it is important to know which virus is causing the problem. Armed with this information, a management strategy can be developed around knowing how the particular virus spreads and what crop plants or weeds in the vicinity may be harboring it. For example, *Tobacco mosaic virus* (TMV) is usually spread mechanically, and can infect tobacco plants. So if TMV is a risk for a particular perennial, it is important to try to determine whether there are other TMV-infected plants nearby. After using tobacco products, nursery workers need to wash their hands with soap and water and dry them thoroughly on a paper towel before touching crops.

Symptoms of *Cucumber mosaic virus* (CMV) on ajuga.

Several companies can test plants for a range of viruses. Some of these companies also sell kits that allow growers to test plants for a specific virus or virus group. A negative result indicates only that the particular virus sought was not present. It is possible that some other virus caused the symptoms or that something other than a virus was responsible for the damage.

A good place to start in managing virus diseases is to purchase only virus-free plants. Avoid using virus-infected plants as stock plants for vegetative propagation. Weed removal in and around plantings is important because weeds may harbor not only viruses, but also insects, mites, or nematodes that may vector viruses. Remove all crop debris from the greenhouse to minimize the chance of virus spread to live plants. Immediately set aside plants with suspicious symptoms and obtain a diagnosis from a plant diagnostic clinic or commercial laboratory. Since some viruses are mechanically transmitted, frequently disinfest tools used for vegetative propagation. When possible and practical, propagate plants by seed rather than vegetatively, since vegetative propagation is far more likely to spread viruses. No chemicals cure a virus-infected plant, nor do any chemicals protect plants from becoming infected. If a virus is carried by aphids, thrips, or mites, the risk of infection can be reduced by monitoring for them and suppressing their numbers when needed—for example, by applying insecticides, miticides, or appropriate biological control measures.

Nematodes

Angular leaf spots, caused by foliar nematode feeding, on anemone.

INTRODUCTION

Nematodes are tiny (1 to 4 mm long), non-segmented worms, many of which are too small to be seen with the naked eye. Plant-parasitic nematodes cause diseases that are often mistaken for cultural problems or fungal root rots. Most plant-feeding nematodes live in the soil and feed on plant roots, but some feed inside leaves or on other above-ground tissues. The plant-parasitic nematodes have a hollow spear-shaped mouthpart called a stylet that is used to penetrate plant cells, inject digestive juices and pull in plant cellular material.

Symptoms and signs

Probably all plants have nematodes feeding on their roots, but generally it takes a large population to cause root dysfunction and decline. An exception would be the root-knot nematode, which can cause stunting with relatively low populations. Because the roots are commonly targeted, often the symptoms seen above-ground are those to be expected when roots are not functioning well: stunting, yellowing, or other indications of nutrient deficiency. In order to determine whether plant parasitic nematode populations are high enough to damage plant health, a soil sample should be submitted to a nematode diagnostic lab; most of these labs are located at universities.

One of the most damaging nematodes, with a wide host range, is the root-knot nematode, *Meloidogyne*. When root knot nematodes feed on the roots, plants may be severely stunted, especially in sandier soils. Direct examination of the roots will show distinctive lumpy swellings ("root knots" or galls) in areas where the tiny female nematodes are feeding. The stem and bulb nematode, *Ditylenchus*, may cause stunting and distortion of the leaves of bulbous plants, and rings within the bulbs themselves.

Another special group of nematodes called "foliar nematodes" (*Aphelenchoides*) may attack above-ground, swimming sinuously up onto the plant while the surface is wet and entering the leaves through the stomates. There they feed within the leaf tissue, mate, and have their offspring, resulting in discolored patches in the invaded leaves. Foliar nematode injury is usually bounded by the major veins in the leaf, so that the patches are rectangular or wedge-shaped. These discolored patches in the leaf may look very much like downy mildew infections, which also tend to be vein bounded. To be sure that symptoms are caused by foliar nematodes, leaves with suspicious discolored patches may be torn into small pieces and placed in lukewarm water in a Petri plate or other small transparent dish. With a little magnification (a dissecting microscope works well), the foliar nematodes may be seen swimming out of the edge of torn lesions after a few minutes to several hours. Lighting the dish from beneath is best for visualizing the nematodes.

Ecology

Plant-feeding nematodes have an egg stage and four progressively larger larval stages before becoming a mature adult that is less than 2 mm long. Some of these nematode species live in the soil and feed on roots (ectoparasites), while others enter the root and feed from within (endoparasites). They may cause injury directly, by feeding on roots, or indirectly, such as by making wounds that help disease-causing fungi to enter roots. Nematodes can also transmit viruses. Certain viruses in the nepovirus group (including *Tobacco rattle virus*, which is common on herbaceous perennials) may be spread from plant to plant through the feeding of *Xiphinema*, *Longidorus* or *Paralongidorus* nematodes. Nematodes move only short distances on their own, but they can be moved along with soil particles by water flow, on cultivation equipment, or on bulbs or other plant materials. Interstate movement of nursery material provides long distance transportation. Root-knot nematodes are becoming more common in the nursery trade.

Management

In field agriculture, highly toxic fumigants have been used for reducing populations of plant-parasitic nematodes in the soil in the past. Some of these materials are now being phased out of use because of their toxicity to the applicator and the danger that they pose to ground water or other aspects of the environment; none are available to a home gardener.

For commercial field nursery production, few products are available for soil application to control nematodes. Check with your local cooperative extension office to learn whether there are chemical control options available in your area. Some systemic insecticides reduce nematode populations and are available for application to container-grown plants in nurseries. Hot water treatment is used to eliminate *Ditylenchus* from dormant narcissus bulbs.

For the home gardener, there are no chemical options. Fortunately, supplemental fertilization and other good cultural care can mask nematode symptoms that would be apparent on plants growing under more stressful conditions. This allows gardeners to tolerate low levels of nematode infestation in their plantings. To avoid the problem, however, always be careful to choose vigorous looking plants when purchasing herbaceous perennials. If plants can be removed from their containers, they can be checked for root knot prior to purchase.

When buying plants, also look carefully for symptoms of foliar nematodes, especially on the older foliage

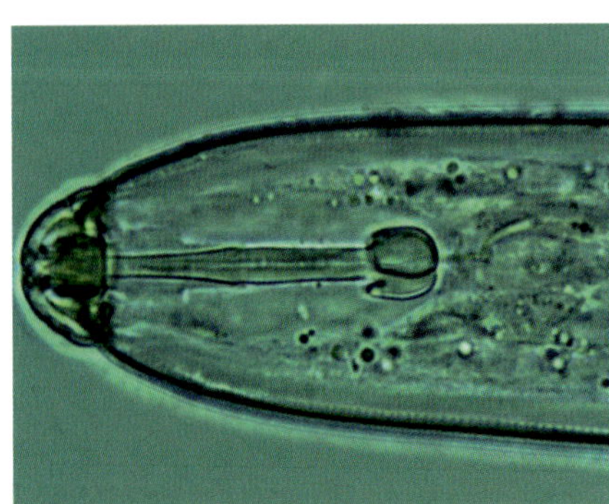

Close-up of a nematode's mouthparts, showing the tapered stylet anchored by an enlarged base.

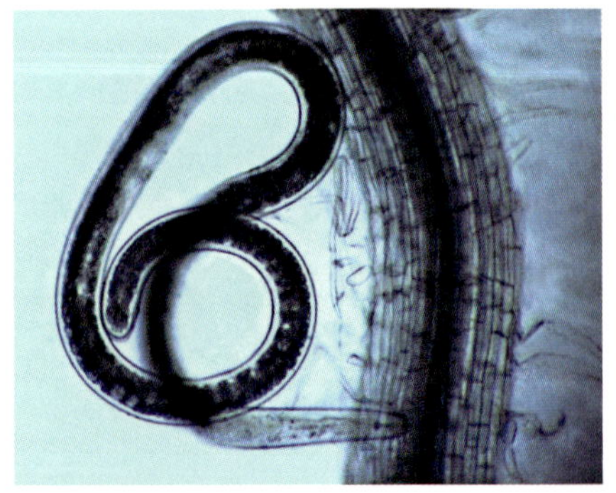

A nematode (coiled) with its head inserted into a root.

Swollen, distorted roots resulting from attack by root knot nematodes.

where the injury first appears. This is especially important for species that are very prone to foliar nematode, such as Japanese anemones. Try to prevent bringing foliar nematodes into your garden on new plant purchases. If discovered later, the best recourse is to hand-pick leaves showing foliar nematode symptoms and to adjust irrigation so as to reduce the amount of time that the leaves sit wet. Growing plants drier in nurseries or gardens will reduce foliar nematode injury.

If root knot nematode has seriously harmed individual plants in an established garden, the plants should be removed carefully, along with the soil clinging to the roots. Be careful not to spread the nematode to areas of the garden that are not already infested. It will be best to replace with a different plant species that is less susceptible to the type of nematode involved. Soil solarization, keeping an area fallow, or planting nematode-suppressive marigold varieties are techniques that can be used to clean up a garden area in which nematode populations have become damagingly high.

Foliar nematode damage on asarum.

Diseases by Host Plant Genus

Acanthus

Acanthaceae

Acanthus, known by the colorful name of bear's breeches, requires rich soils that are light and well-drained, and sunny locations in the garden. Plants can be killed by overly wet conditions during winter. Mulches can be important for helping winter survival in the north.

Plants in the genus *Acanthus* are rarely troubled by diseases, although slugs and snails may visit them in spite of their formidable foliage. *Acanthus* species are susceptible to root knot nematode (*Meloidogyne*), which can cause small bump-like galls on the roots. Powdery mildew caused by a fungus and shoot proliferation caused by the bacterium *Rhodococcus fascians* (previously known as *Corynebacterium fascians*) have also been reported on *Acanthus* species but appear to be rare.

Distortion and adventitious bud development on the leaf of an acanthus infected with *Rhodococcus fascians*.

Clump of adventitious buds developing on acanthus stem infected with *Rhodococcus fascians*.

Achillea

Asteraceae/Compositae

Achillea, or yarrow, is a popular and reliably performing perennial. Many varieties and hybrids are available; the flowers vary in color from species to species. These plants appreciate sunny or partly shaded sites, and tolerate poor soils as long as drainage is good. If night temperatures are too warm the stems will topple, and some flowers will fade in hot weather. Dividing plants and replanting will rejuvenate them if they have ceased to flower well.

Yarrows are generally disease-free in favorable climates, but insects can cause some symptoms that may be mistaken for diseases. The foliage can be browned by feeding injury and droppings from lacebugs and thrips. White foamy spittlebug masses are sometimes observed on stems. Few problems occur on yarrow during production and most can be controlled through judicious use of cultural and chemical controls.

A rust disease caused by *Puccinia millefolii* will cause chocolate brown bumps to appear on the leaves. Wet summer conditions favor this fungal disease, which is most often seen in the western and southwestern states. In wetter climates, *Botrytis cinerea* can attack flowers or stems, or leaf spots may occur after infection by species of the fungi *Alternaria*, *Cercospora*, *Leptosphaeria*, *Pleospora*, or *Septoria*. All of these are kept in check by careful attention to irrigation practices. Powdery mildew, crown gall and root knot nematode are also possibilities. Yarrows are sometimes troubled by lower stem cankers caused by *Rhizoctonia solani*, especially if mulched too deeply. Pythium root rot and Phymatotrichopsis root rot can also occur. Avoid excessive overhead irrigation of yarrow to prevent foliar and root rot disease problems.

Rust pustules on yarrow may be fairly inconspicuous because the leaves are so narrow.

Close-up views of chocolate-brown rust pustules on yarrow.

Aconitum

Ranunculaceae

Aconitum, or monkshood, is a genus of plants with an aura of mystery about them due to their hooded flowers and the poisonous character of both the roots and above-ground parts. Monkshoods are not appropriate plants to grow adjacent to root crops in the vegetable garden—but these flowers have an unusual advantage in that they are left alone by deer, and their curiously hooded form is enchanting. The two-petaled flowers help to separate them from delphiniums, which have a similar leaf. They tolerate some later afternoon shade. Dry to moist (but not soggy) soils are acceptable, and richer soils are preferred. Aconitums are better adapted to growing in the northern United States or higher elevations in the south, as cool nights are best for their health. Winter mulching with a layer of leaves helps survival in the north. If plants become crowded, their flowering will decline, so make divisions and replant every few years, in the fall. Plants may need a season to fully recover from transplanting.

Aconitums are quite susceptible to Verticillium wilt caused by *Verticillium albo-atrum*. The affected plants will perform poorly in the garden—their leaves will turn brown and dry, and flowering will be reduced. Cutting across the stem of an affected plant will reveal dark discoloration of the vascular bundles (water conducting tissues). If this disease is detected, do not try to replant monkshood in the same part of the garden, as the microsclerotia of the fungus will remain in the soil for several years to attack new specimens.

Monkshoods are susceptible to many of the same diseases as their close relatives, delphiniums. These include black leaf spots caused by *Pseudomonas* bacteria and the white colonies of moldy growth on leaves caused by the powdery mildew fungi. Foliar nematodes (*Aphelenchoides*) may cause discolored areas in the leaves. Stem rots may be caused by the fungi *Sclerotinia sclerotiorum*, *Sclerotium rolfsii*, *S. delphinii* and *Rhizoctonia solani*, and root rot caused by *Phymatotrichopsis omnivora* can cause problems in the extreme southern United States. Root knot nematode, rust, smut and downy mildew diseases have also been reported for *Aconitum* species. Aconitum is susceptible to *Impatiens necrotic spot virus* (INSV), a thrips-borne virus that can cause a variety of symptoms including leaf spots and mottling of foliage, as well as *Cucumber mosaic virus* (CMV), which is spead by aphids and can cause mottling of leaves.

Swollen areas on the roots of monkshood are symptomatic of infection by root knot nematode, *Meloidogyne hapla*.

Foliar nematode infestation (*Aphelenchoides* sp.) has caused the blackened, aborted sprout on this monkshood transplant.

Mottling of monkshood leaves, caused by CMV.

The underground stems of monkshood show blackening of the vascular system due to Verticillium wilt.

Actaea

Ranunculaceae

Actaea (previously called *Cimicifuga*), known as cohosh or baneberry, features some poisonous parts—the roots and the handsome berries. These plants appreciate a woodland environment, so they should be grown in rich, moist soils with plenty of organic matter, in shade. Similarly to astilbe, actaea will be subject to injury from dry periods in the summer and will need supplemental irrigation in many climates.

Few diseases are known for these plants. Some fungal leaf spots caused by species of *Ascochyta*, *Ramularia*, and *Phyllosticta* have been reported. Leaf spotting may also be caused by foliar nematodes, which cause brown patches between the veins when they feed within the leaf tissue. *Actaea* spp. are susceptible to some rust fungi, including *Puccinia recondita* that has certain grasses as its alternate hosts. Smut diseases caused by species of *Urocystis* have also been reported. These foliar problems along with foliar nematode injury will be kept in check by minimizing periods of leaf wetness.

Vein-limited necrotic areas in actaea leaves infested with foliar nematodes (*Aphelenchoides*).

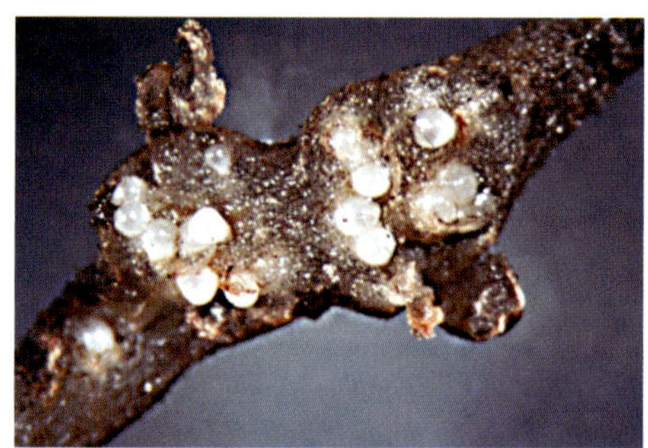

The globose white structures in the swellings on this actaea root are the swollen bodies of root knot nematode females (*Meloidogyne*).

Yellow mottling that indicates a possible virus infection on actaea.

A fungal leaf spot caused by an *Ascochyta* species on actaea.

Aegopodium

Apiaceae/Umbelliferae

The genus *Aegopodium* contains a single species, *A. podagraria*, whose common names include snow on the mountain and bishop's weed.

The variegated form *A. podagraria* 'Variegatum' is not as vigorous as the species and appears to be especially susceptible to Septoria leaf spot, which is the most important disease on aegopodium. This fungus infection causes irregular brown spots or marginal scorched areas on leaves, and the dead areas usually are misinterpreted as being a consequence of sunscald. Close examination, however, will show the tiny rounded spore bodies (called pycnidia) on the upper surface of the dead spots. Long, thin, multiseptate spores ooze out from the pycnidia and will easily splash from plant to plant in massed groundcover plantings or in closely-spaced nursery containers. To minimize injury from Septoria leaf spot, overhead irrigation should be timed for early in the day so that the foliage has time to dry before nightfall. Fungicides may be used in nurseries to reduce the incidence of these fungal infections. This is the only disease problem noted in the United States. Powdery mildew, downy mildew and a few leaf spots are reported from Central Europe, the United Kingdom or Asia.

Large scorched areas may develop where multiple Septoria leaf spots have coalesced on aegopodium leaves.

Fungal leaf spots caused by *Septoria* sp. are common on aegopodium.

Agapanthus

Alliaceae/Liliaceae

Agapanthus or African-lily species vary in their hardiness; many are somewhat tender. They are often grown in tubs to allow movement into more protected environments during the winter. Blooming plants need large quantities of water. When dormant, the plants can tolerate light frost.

During production in the warmer portions of the United States (Florida, Texas and California especially), *Agapanthus* species are subject to a variety of crown rots caused by *Pythium*, *Phytophthora* and *Pectobacterium* species. They are best controlled by carefully examining plants on a routine basis and discarding those with wilting or basal discoloration. There are no effective bactericides for soft rot, which is the disease caused by *Pectobacterium*. In contrast, several fungicides, used in rotation, can be helpful in preventing both Pythium and Phytophthora root rot. Additionally, avoiding overhead irrigation, exposure to rainfall and overwatering in general will reduce losses due to these two water molds as well as *Pectobacterium* or other soft rot bacteria.

In the garden, agapanthus is not commonly affected by diseases, but Botrytis blight is a possibility if water sits on the foliage. A Phoma leaf spot has been reported from Hawaii on *A. umbellatus* and *A. africanus*. The plants are also susceptible to *Ornithogalum mosaic virus* (OrMV), and infection with *Tomato spotted wilt virus* (TSWV) has been reported from Australia. Virus symptoms are common in some parts of the United States but specific viruses have often not been identified. If the symptoms (mottling, ring spots or mosaic, for example) are associated with reduced vigor, you can remove the plant and replace with one that does not have any virus symptoms.

Reduced root system and softened and blackened root tips caused by Pythium root rot on agapanthus.

These black stem lesions on an agapanthus stem are caused by a *Botrytis* species.

Dead, dry collapsed areas on agapanthus caused by exposure to an herbicide.

The rot at the crown of this agapanthus plant is due to infection by a *Phytophthora* species.

Agastache

Lamiaceae/Labiatae

Agastache, or giant hyssop, contains very aromatic members of the mint family; many species have scented foliage, and the flowers may also be very appealing. These plants are not very hardy, so they are sometimes grown as annuals. They need moist but well-drained soil in full sun or light shade.

Several species and hybrids of agastache are susceptible to downy mildews caused by *Peronospora* species. Vein-limited pale patches that turn brown over time will develop on the leaves of plants that are infected by the same downy mildew that infects coleus. This pathogen also infects some other members of the mint family, such as basil. Whether you are growing agastache in commerical production areas or in the garden, keep downy mildew problems minimized by watering early in the day. This reduces humidity around the plants and thus helps reduce downy mildew in the same way that spacing them to allow good air movement will help. Several fungicides are very effective in preventing downy mildew. If you had this problem in previous seasons, preventive fungicide applications may be needed in the nursery. Gardeners may need to experiment to find varieties that are less threatened by downy mildew.

Several fungal leaf spots are also possible on agastaches, including spots due to *Ramularia*, *Septoria* and *Ascochyta* species. Powdery mildew has been reported from the western United States, and there is also a possibility of a rust disease caused by *Puccinia hyssopi*. The telial (overwintering) stage of the rust is formed on agastache. Gather and remove plant debris from the garden in the fall to reduce the overwintering of this rust fungus.

Yellow mottling and patterns formed by yellow lines in the foliage have been associated with infection of agastache by an unknown virus. Symptoms caused by *Impatiens necrotic spot virus* (INSV) have also been encountered during greenhouse production; these include a light yellow mottling as well as necrotic spots and blotches in the leaves. Virus infected plants should be disposed of as soon as symptoms are noticed. *Agastache* spp. are also hosts of Verticillium wilt.

Two different kinds of unidentified virus particles were found in the sap of this agastache.

Vein-limited patches in agastache here indicate downy mildew infection; very similar symptoms could be caused by foliar nematodes.

Brown spots and yellow mottling may indicate *Impatiens necrotic spot virus* (INSV) in agastache.

Ajuga

Lamiaceae/Labiatae

Ajuga, bugleweed species, will perform best with frequent dividing and replanting. These stoloniferous plants do well as ground covers, and are easy to grow under a wide range of conditions.

Powdery mildew may develop on ajugas if humidity is high, forming a thin white coating across the surface, or whitish patches scattered on upper leaf surfaces. Frequent overhead watering, high humidity and poor air circulation may encourage powdery mildew to develop.

In hot climates, the fast-spreading common bugleweed, *Ajuga reptans,* is prone to crown rot caused by *Sclerotium rolfsii* and *S. delphinii.* The crown rot will be noticed when large areas of plants turn brown. Close examination of plants in these patches will usually reveal the tell-tale sclerotia of the fungus. These are little round bodies that look remarkably like mustard seeds; you'll find them clustered on stems near the soil. The sclerotia are often accompanied by wefts of white mycelium of the fungus at the base of the stems. Sclerotium crown rots are most likely to be troublesome in humid southern climates, hence the common name of the disease, "Southern blight". Frequent division to prevent overcrowding will help to minimize the effects of Southern blight by improving air movement around plants. If an area blighted by this fungus is found, remove all plant material (taking care not to drop bits of infected plants and soil onto nearby healthy ajugas). Replanting the bare spots early the next spring with divisions from healthy areas and possibly drenching with a fungicide might help new plants get started. *Sclerotium* has a very wide host range, so be careful not to spread the fungus within a nursery, or to other plants in a garden.

The other likely crown rot fungus on common bugleweed, *Rhizoctonia solani,* also has a wide host range. Rhizoctonia crown rot lacks the conspicuous white wefts of mycelium and round tan sclerotia that are characteristic of the other fungus. Instead, *Rhizoctonia solani* has relatively thin, cobwebby brown mycelium and its sclerotia are small, brown, and inconspicuous. *Fusarium oxysporum* and *Fusarium solani* have been reported to attack ajuga stems occasionally, and *Phoma* species can also cause crown rot. Fungicides are available to limit the spread of these fungi during nursery production.

Other fungi can cause leaf spots on ajugas, but these are generally minor problems. Some of the fungi reported to cause leaf spots are species of *Alternaria*, *Cercospora*, *Corynespora*, *Colletotrichum* and *Myrothecium*.

Root knot nematode, *Meloidogyne incognita*, can also attack ajugas, causing roots to be stunted and deformed by

Alfalfa mosaic virus (AMV) caused the mottling and distortion on the young leaves of this ajuga plant.

Yellow rings on an older leaf of ajuga due to *Alfalfa mosaic virus* (AMV).

Cercospora leaf spot on ajuga.

Cucumber mosaic virus (CMV) caused leaf mottling on ajuga.

small swellings (galls). Above-ground symptoms include unthrifty growth and a look of nutrient deficiency. Plants attacked by *Pythium* will also look small and unthrifty, but will show softening and discoloration of the roots.

Ajuga frequently shows symptoms of virus infection in the landscape. These include ring spots, mottling, line patterns, yellow spots and yellowing along the margins of the leaves. Some plants are infected but symptom-free. Symptoms alone cannot be used to identify the viruses, as they overlap with one another. A survey at Ohio State in 2000 found *Alfalfa mosaic virus* (AMV), *Tobacco streak virus* (TSV) and *Cucumber mosaic virus* (CMV) in 11% of 356 samples examined from commercial nurseries. These viruses might be seed-borne in ajuga and certainly would be spread when plants are propagated by division or taking cuttings. Both CMV and AMV are spread by dozens of different aphid species, while TSV is transmitted by two thrips species. It is important to remove from commercial production areas any ajuga plants that show virus-like symptoms.

Alcea

Malvaceae

The genus *Alcea,* or hollyhock, contains plants that are biennials or short-lived perennials. The proper site for hollyhock is a sunny spot with good soil drainage and air circulation. These conditons will help to reduce the impact of the classic disease that is otherwise relentless on *Alcea rosea*: hollyhock rust. The fungus causing this disease, *Puccinia malvacearum*, forms small yellow-orange spots in the leaves. Opposite these spots, on the leaf undersurface, are round orange-brown bumps that protrude from the leaf. This rust does not have an alternate host, but the common mallow weed, *Malva rotundifolia*, can harbor the rust fungus and supply spores to your hollyhocks. Some of the modern varieties are less prone to hollyhock rust than those grown in the past. The species *A. rugosa*, the Russian hollyhock, has a yellow flower and less susceptibility to rust. *Alcea ficifolia*, fig-leaf hollyhock, is also relatively resistant. Both of these resistant species can be grown from seed. Gather and destroy all plant debris in the fall, because this is where the fungus overwinters, in the form of teliospores on leaves and stems. In production, hollyhocks with rust should be removed and destroyed while the remainder can be sprayed with one of many effective rust fungicides. Be sure to add a wetting agent to the fungicide spray to improve penetration of the rust pustule and increase efficacy.

Numerous other leaf-parasitic fungi may cause blemishes on hollyhocks. The anthracnose caused by *Colletotrichum malvarum* is widespread in the eastern United States. Other leaf fungi on hollyhocks include *Ascochyta althaeina*, *Cercospora althacina* and *C. kellermanii*, plus *Myrothecium roridum* and *Septoria malvicola*. All of these will cause leaf spotting; these spots are neither yellow nor swollen, so they should not be easily confused with hollyhock rust. Powdery mildew will form whitish patches, mainly on the upper leaf surface.

Rhizoctonia solani, Sclerotinia sclerotiorum and *Sclerotium rolfsii* are three fungi that may cause cankers at the base of hollyhock stems. Crown rot may also be due to the water mold *Phytophthora megasperma*. Avoiding excessive soil moisture will be helpful for preventing attack by *Phytophthora*. Keeping mulch away from the stem will reduce humidity near the stem surface and help to manage problems from stem-attacking fungi. If plants wilt suddenly and fungal mycelium or a dead, brown cankered area is seen, remove the dead plant carefully along with the soil in the immediate vicinity, to reduce the chance that the problem will spread to nearby plants. To avoid escalating problems with crown rot, stem rot or root rot, never reuse potting media during production of hollyhock. Reuse only containers that have been washed and disinfested.

Vein-limited spots indicate Cercospora leaf spot on hollyhock.

Depressed yellow to orange spots on the upper surface and brownish bumps on the undersurface indicate hollyhock rust.

Phyllosticta leaf spot takes the form of large zonate lesions on hollyhock.

Alchemilla

Rosaceae

Alchemilla, or lady's mantle, is a genus of plants grown primarily for their attractive pale green pubescent foliage, which makes it an attractive edging plant. The flowers of the popular *Alchemilla mollis* are green or yellow, and lack petals. They prefer a moist soil in a sunny site. *Alchemilla* thrive in Northern gardens, but can take more heat if provided partial shade. The foliage can be renewed by cutting plants down to the ground after flowering.

Alchemilla mollis is generally trouble-free. When there is heavy rainfall or leaves are wet for long periods following overhead irrigation, large brown leaf spots caused by *Botrytis cinerea* may develop. If this happens, merely remove spotted leaves and adjust watering practices. A few other leaf problems occur occasionally, including Colletotrichum and Septoria leaf spot, but these can be minimized with careful watering practices.

Botrytis blight will appear on lady's mantle in rainy weather or under excessive overhead irrigation.

Under wet conditions, lady's mantle may show leaf spotting due to *Colletotrichum*, an anthracnose fungus.

Allium

Alliaceae/Liliaceae

Allium is the genus of onions, some of which are wonderfully decorative perennials. The flowering ornamental plants in this genus conveniently do not give off any onion-like aroma unless they are trodden upon or roughly handled. With some of the early-flowering ones, the leaves naturally die back and disappear either before or just after flowering. Exceptions to this are the Turkistan onion, *A. karataviense*, and the Japanese onion, *A. thunbergii*, both of which maintain their foliage. Alliums need sunny, well-drained locations in the garden. Their longevity is reduced by humid rainy climates or by planting in overly heavy soils. Persian onions, *A. aflatunense*, for example, are prone to root and bulb rot under these conditions. To reduce the chance of winter kill, alliums should be mulched heavily in the more northerly parts of their range.

Few if any diseases have been observed on ornamental onions, but their edible relatives are host to a wide range of disease organisms. Downy mildew caused by *Peronospora destructor*, foliage diseases caused by *Botrytis cinerea*, *B. allii*, and *B. squamosa*, and smut caused by *Urocystis cepulae* are some of the most common problems on onions used for food, and some of these might become problematic on ornamental alliums in the future. For now, bulb rots associated with winter chill and poor drainage are the greatest challenge to the gardener. Production challenges include Fusarium and Sclerotium bulb rots. Any plants with symptoms of foliar nematode (*Aphelenchoides*) and stem and bulb nematode (*Ditylenchus*) should be discarded.

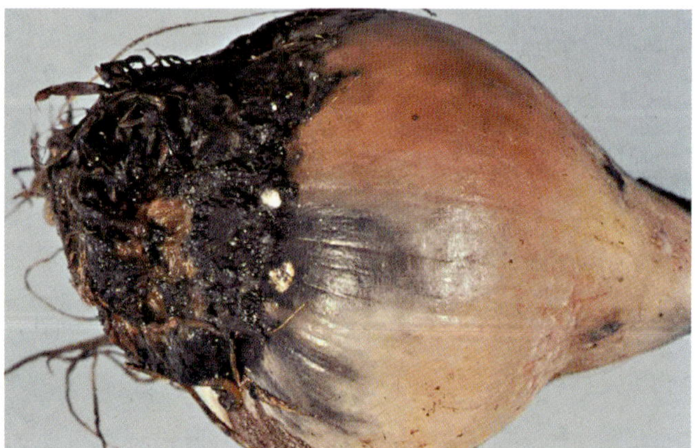

Sclerotium cepivorum has blackened the base of this allium bulb.

Dead ornamental allium leaves covered with sclerotia of *Sclerotium rolfsii*. The larger, bluish structures in the picture are fertilizer granules.

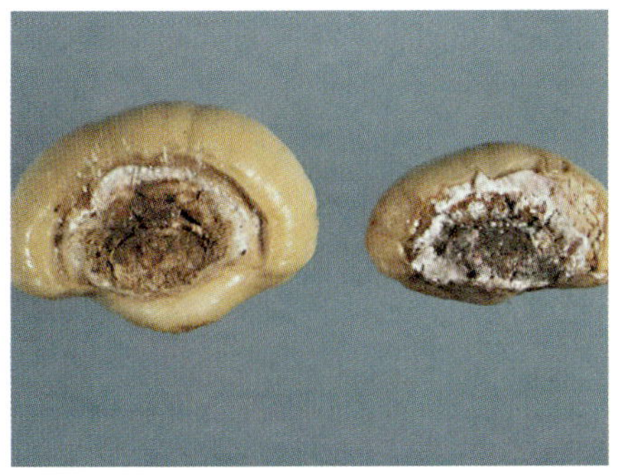

Bulb rot due to *Fusarium oxysporum* on an ornamental allium.

Allium bulb discoloration due to the stem and bulb nematode, *Ditylenchus dipsaci.*

Browning at the basal plate of allium bulbs caused by *Aphelenchoides subtenuis.*

Alstroemeria

Alstroemeriaceae/Liliaceae

Alstroemeria, or Peruvian lily, is a genus of plants that are native to Brazil and Chile. Although more familiar as items in restaurant bud vases, alstroemerias can also be used in landscapes. The roots of *Alstroemeria* spp. are easily injured during transplanting, which can lead to a delay in flowering. Some tolerate partial shade; others fare best in full sun. Moist soil with plenty of organic matter is desirable for alstroemerias.

Alstroemerias are susceptible to Pythium root rot under conditions of poor soil drainage. They are also commonly hosts of *Alstroemeria mosaic virus* (AlMV), which may cause flower break as well as mosaic or yellowing of leaves. This virus can be spread from plant to plant by the green peach aphid, *Myzus persicae*, but it is not spread by handling or through seed. *Impatiens necrotic spot virus* (INSV) and *Tomato spotted wilt virus* (TSWV) have also been reported from alstroemeria in North America, and have been associated with mosaic and mottle of leaves and stem canker symptoms. Yellowing of lower leaves has been found to be due to *Sclerotium rolfsii* attack belowground, rotting the rhizomes. A Canadian report mentions a root rot complex of *Fusarium* and *Rhizoctonia* species plus *Pythium ultimum* on plants that were wilting or showed yellowing leaves or browning leaf margins. When used preventively, fungicides may be helpful in reducing some crown and root diseases on alstroemeria in container production.

Pythium root rot led to stem dieback aboveground in this alstroemeria.

The mottled and necrotic areas in this alstroemeria are symptoms of *Impatiens necrotic spot virus* (INSV) infection.

Alstroemeria mosaic virus caused white patches on the flower at left, and mottling on the leaf above.

Alternaria leaf spot on alstroemeria has tan lesions with purple rims.

Fusarium root rot on this alstroemeria has caused the roots to become watersoaked and softened.

Sclerotia ranging from tan to brown to dark purplish brown may be used to identify crown rot due to *Sclerotium rolfsii* on alstroemeria.

Anemone

Ranunculaceae

Anemone, or windflower, is a diverse genus with various plant forms and flowering times. *A. nemorosa* and *A. blanda* are delicate looking spring-blooming perennials. For these spring anemones, occasional dividing keeps them flowering, and flowering will be best in light shade, in soil with plenty of organic matter and adequate moisture that is well drained. The poppy anemone, *A. coronaria*, is often used as a cut flower. Its flowering will decrease after a few years, so gardens should be replanted with new tubers. Japanese anemones, *Anemone* ×*hybrida*, come into their long flowering in summer and fall. They may take a year to become established after planting out from containers. With sufficient moisture they do well in either sun or shade, but may do best in well drained soils and shade. Some varieties of Japanese anemone can become invasive.

Native anemones may be affected by numerous leaf spotting fungi (species of *Cercospora*, *Didymaria*, *Phleospora*, *Phyllosticta*, *Ramularia* and *Septoria*). There are also rust diseases caused by six species of *Puccinia*, some of which have reeds or grasses as alternate hosts, and five species of *Tranzschelia*. Tranzschelia rusts have been noted in both the southern and western United States; *T. discolor* has *Prunus* spp. as alternate hosts. Two smuts (*Urocystis* spp.) and one white smut (*Entyloma ranunculi*) may affect anemone. Anemone leaves can sometimes show leaf galls caused by *Synchytrium anemones*. Both powdery and downy mildew can also be found in some locales. In spite of these possibilities, it is still very likely that you will not observe any foliar disease problems on the early anemones in your garden. Other kinds of problems are also rare on these plants. *Anemone*

Botrytis stem rot on anemone.

Petal break in anemone petals due to virus infection.

spp. have been reported as hosts for aster yellows, a phytoplasma disease (spread by leafhoppers) that will cause pale foliage and peculiar-looking green flowers. Root knot nematode (*Meloidogyne*) can infest the roots and cause galls which may lead to a stunting of the above-ground parts of the plant.

Japanese anemones are prone to foliar nematode infestation (*Aphelenchoides fragariae* and *A. ritzemabosi*), and this problem is widely disseminated in horticultural trade via infected plants. The nematodes cause yellow vein-limited patches at first, and these darken to ugly brown lesions with time. Very strict sanitation programs are needed in nurseries to control foliar nematode in Japanese anemones. Overhead irrigation can splash nematodes from the soil up onto foliage as well as increase leaf-to-leaf spread. Gardeners should avoid purchasing Japanese anemones showing any yellow or brown discolored patches in the foliage, and promptly remove leaves that develop symptoms. Areas where infested Japanese anemones have been grown in nurseries or landscapes will remain contaminated. Because foliar nematodes have a wide host range, subsequent crops brought into the area are likely to become infected. Avoid following Japanese anemones with other highly susceptible plants, in either garden or nursery. Nurseries should try to obtain nematode-free stock of anemones, and carry varieties that are not highly susceptible to foliar nematode.

Anemone coronaria produced as a cut flower in greenhouses is very prone to Botrytis crown rot. This plant is also susceptible to an anthracnose disease caused by *Gloeosporium* sp., which causes leaf spotting as well as a curious twisting of leaves and pedicels. A wet looking, orange coating of spores can appear on spots under humid conditions. Reduction of leaf wetness and splashing will help to manage this problem. Greenhouse-grown anemones are very susceptible to *Impatiens necrotic spot virus* (INSV); infection may cause yellow mottling or brown spotting on the leaves.

A combination of bacterial leaf spot and foliar nematode infestation on Japanese anemone.

Vein-limited brown leaf spots caused by foliar nematode on anemone foliage.

Early symptoms of foliar nematode injury on Japanese anemone.

Necrotic lesion on *Anemone coronaria* leaf due to *Impatiens necrotic spot virus* (INSV).

Elaborate line patterns in a leaf of Japanese anemone infected with *Tobacco rattle virus* (TRV).

Aquilegia

Ranunculaceae

Aquilegia, or columbine, is a genus of delicate plants with flowers ranging from pastels to very vivid hues. The name comes from the spurs at the back of the petals, which are reminiscent of eagle claws (The Latin word for eagle is "aquila"). The best cultural conditions for aquilegia are those that mimic moist woodland slopes in early spring: light shade and a light sandy soil with abundant moisture and good drainage. Mature plants can tolerate heavier soils than young seedlings. Aquilegias will be prone to crown and root rots in overly heavy soils.

Septoria leaf spot on columbine.

The most conspicuous problem of aquilegias is leafminer: the feeding of the larvae of this fly creates meandering pale green trails in the foliage that are seen wherever the plant is grown. Also, particularly if conditions are moist in nursery or garden, either purplish bruised spots or white coatings caused by the powdery mildew fungus may appear on the leaves. This powdery mildew will usually not appear until after flowering, and does not severely weaken the plant. Hence, it is possible to tolerate the presence of some powdery mildew on aquilegia in the garden. In production, however, powdery mildew can be controlled with either preventative or curative treatments of a wide variety of fungicides.

Less commonly other fungal infections may appear, in the form of round or oval leaf spots. The most common of these is Alternaria leaf spot, which appears as light brown oval spots a quarter-inch or more in diameter, often with a purple rim. The production of dark brown many-celled spores on tiny stalks at the center of the lesion can make the center of the lesions look "dirty" to the naked eye, but the fungus sporulation will be easier to see with a hand lens. Other fungal leaf spots on

Unidentified bacterial leaf spots on columbine.

Xanthomonas leaf spot on columbine.

aquilegia are usually caused by species of *Ascochyta*, *Cercospora*, *Haplobasidium* or *Septoria*. During spring greenhouse production, columbine is especially susceptible to Botrytis blight, which attacks the petioles and crown of the plant as well as causing leaf spots. Bacterial leaf spot is also seen occasionally on aquilegia: both *Xanthomonas* and *Pseudomonas* spp. have been implicated as pathogens. A rust fungus (*Puccinia recondita*) with grasses as alternate hosts and a smut (*Urocystis sorosporioides*) are reported on columbine, but neither is common in gardens or in production. To deter fungal and bacterial diseases, water early in the day to allow foliage to dry as soon as possible and maintain sufficient spacing between plants. In greenhouse production, use fans to improve air circulation and vent and heat at sunset to reduce humidity around plants.

Foliar nematode may infest aquilegia, particularly under wet, shaded production conditions in nurseries, where plants of one species are closely gathered together. The nematodes live in the leaf and feed on plant cells, with the result that patches within the leaf turn yellow, then dark brown. Nematodes initially gain entrance to leaves through the stomata and are limited in their ability to move across major veins, hence the dead areas are limited by veins and have straight sides rather than rounded edges. There are no chemical controls for foliar nematode. Removing infested lower leaves and irrigating so as to reduce leaf wetness are the principal control measures. Root knot nematode (*Meloidogyne*) may also infect columbine, causing small galls (swellings) on the roots. Replanting with species not highly susceptible to root knot is the only control practice available in the garden. Nurseries should discard any plants found to be infested.

Mosaic patterns (alternating patches of yellow with the normal green) may appear in columbine foliage due to infection by a virus, sometimes *Cucumber mosaic virus* (CMV). Discard entire plants if these symptoms are noted, as certain aphid species are able to spread CMV to other plants in your garden. Another potential virus problem on aquilegia is *Impatiens necrotic spot virus* (INSV), which is vectored by the Western flower thrips and has a wide host range. Various symptoms including leaf mottling and stunting can indicate INSV infection.

Root problems that might cause wilting or stunting on columbine include Phymatotrichopsis root rot (almost exclusively in certain southern states), black root rot (*Thielaviopsis basicola*) and Pythium root rot. Species of the fungi *Sclerotinia*, *Rhizoctonia* and *Phoma* have caused stem rots. *Sclerotium rolfsii* may also cause rotting at the crown. If crown rots have been a problem, cultivate the soil around columbines so that it will dry rapidly, rather than mulching up to the base of the plant. Replace the soil in the immediate area if you need to remove a plant that has died from crown rot.

Foliar nematode causes dark wedges of brown discoloration in columbine leaves.

Leaf miner trails in columbine leaves are sometimes mistaken for disease symptoms.

The yellow mottle on these columbine leaves is due to *Cucumber mosaic virus* (CMV).

Alternaria leaf spot on columbine.

This Botrytis blight on columbine began where a piece of dead petal tissue landed on the leaf.

Powdery mildew on columbine is most common in very moist environments.

Injury to columbine leaves following a copper fungicide application.

Arabis

Brassicaceae

Arabis or rockcress is a genus of common rock-garden perennials. These plants perform better in northern climates: the center of clumps often melts out in the southern United States. They should be planted in full sun and provided with good drainage. They will do well even in poor soils.

Given appropriate growing conditions, arabis is usually trouble-free in garden and nursery. In the absence of good drainage, however, *Arabis* spp. are prone to Pythium root rot. *Rhizoctonia solani* may cause damping-off of young seedlings. Poor aeration can increase the opportunity for Botrytis blight of flowers, foliage and stems. High humidity can also encourage the development of downy mildew caused by the fungus-like organism *Peronospora*. Arabis is susceptible to the same downy mildew that affects cabbage, a close relative. Leaves affected by downy mildew may show yellow or purplish areas, with a downy mass of sporulation produced on the undersurface of the leaf. Leaf galls are caused by a *Synchytrium* species. Arabis is also susceptible to white rust caused by one of two *Albugo* species. This disease causes white blisters to form on leaves or stems. Leaf spots on arabis may be caused by a variety of fungi including some species of *Cercospora*, *Mycosphaerella*, *Phyllosticta* and *Septoria*. Rust-infected arabis leaves show reddish-brown pustules containing spores of one of several *Puccinia* spp. that affect plants in this genus.

Argyranthemum

Asteraceae

Argyranthemum, or dill daisy, is a group of tender perennials or subshrubs including *A. frutescens*, marguerite daisy, which is used as a cut flower, flowering potted plant and bedding plant. In the northern United States these plants are often treated as annuals, as they are hardy only to Zones 10 or 11. They should be grown in well-drained but moist soils in full sun to light shade, with protection from wind. Deadheading will bring on more blooms.

Argyranthemums are susceptible to downy mildew caused by *Peronospora radii*. The infected plants often show systemic symptoms that may be hard to identify at first. These are most likely to be seen in greenhouse production in the spring. The younger foliage will become chlorotic and stunted, in some cases turning brown and dying. Some leaves will show yellow blotches that represent more localized infections. Because the leaves are narrow, only close inspection with a hand lens or microscope will allow discovery of the downy mildew sporulation on the undersurface of the leaves. If this problem is encountered during production, discard the infected plants and use systemic and contact fungicides in rotation to protect the rest of the crop. In the garden, remove plants showing downy mildew symptoms.

Crown gall, a disease caused by the bacterium *Agrobacterium tumefaciens*, is managed primarily by careful sanitation. The galls are formed at the base of the stem or on the roots. They appear as rounded swollen lumps, often lighter in color than the stem tissue. The pathogen has a wide host range, so if crown gall appears on argyranthemums during greenhouse or nursery production, discard the plants immediately and do not reuse the growing mix or containers. Argyranthemum is also attacked by the bacterium *Rhodococcus fascians*.

Rarely, a Cercospora leaf spot may be seen on argyranthemum foliage. Spores that develop on the round brown spots can be splashed to new plants by overhead irrigation. Remove diseased leaves if practical, and cut back on the frequency or duration of irrigation if either is excessive. The only other occasional foliar problem on argyranthemum is rust. Brown spore pustules develop on the undersurface of rust-infected leaves.

Root diseases of argyranthemum include Pythium root rot, which is most likely in poorly drained soils or in containers that are irrigated more than necessary. Avoid mulching up to the stem in the garden to prevent losses from Rhizoctonia crown rot or other fungal stem diseases.

Downy mildew on argyranthemums grown as potted plants causes symptoms similar to those seen on cut flowers.

Swollen tissue at the root crown of argyranthemums indicates crown gall.

Cercospora leaf spot on argyranthemum.

Yellowing and stunting of upper foliage of argyranthemums caused by downy mildew.

Rust on argyranthemum.

Pythium root rot may cause stunting or death of argyranthemum.

Arisaema

Araceae

Arisaema, or Jack-in-the-pulpit, is ideal for cool, moist, shady gardens with lots of humus in the soil. In moist, shaded sites, foliage will persist longer than in open areas. Sometimes late frost can catch *A. sikokianum* (a Japanese native) in southern gardens because it begins growth so early in the spring. Most species go into dormancy after the spring flowering period.

Jack-in-the-pulpits have a susceptibility to one significant problem, a rust disease. The rust fungus *Uromyces caladii* may affect one or more individuals in a collection, but rarely affects all of the plants. In its most severe form, the rust infection will cause distortion of the leaves and spathe tissue, but at a minimum it will produce a coating of tiny bumps that are the spore pustules of the rust fungus. These become orange in color as the fungus produces its spores. Avoid purchasing plants that show signs of this disease. All of this rust's spore stages, including those for spreading during the season and for overwintering, are produced on Jack-in-the-pulpit—no alternate host is involved.

A leaf and stalk blight caused by *Streptotinia arisaemae* (syn. *Botrytis streptothrix*) has also been reported on Jack-in-the-pulpit from several states. Avoiding excessive overhead irrigation will be helpful for keeping rust and leaf and stalk blight from spreading extensively.

The underside of a Jack-in-the-pulpit leaf may be covered with rust spore pustules.

Closeup of rust spore pustules on Jack-in-the-pulpit, showing the masses of bright orange spores.

Armeria

Plumbaginaceae

Armeria species include some popular rock-garden plants known as sea-thrift that are easy to grow. Full sun locations are ideal for armeria except in hot southern climates, where a bit of afternoon shade is beneficial. *Armeria maritima* is salt-tolerant.

Three rusts, *Uromyces limonii*, *U. limonii* var. *armeriae*, and *U. armeriae*, have been reported on species of *Armeria* from California, and *Uromyces armeriae* and various subspecies have been reported from Oregon and Canada. Leaf spot pathogens in the genera *Pleospora*, *Cercospora* and *Colletotrichum* have been noted on *Armeria* as well. Otherwise this plant is known to have no problems in North America other than Rhizoctonia and Phytophthora crown rots. Spacing plants adequately and keeping mulch away from their base may help minimize losses from crown rot diseases. None of these diseases is common during production of sea-thrift.

Winter injury to armeria is most likely when plants sit wet during the winter.

Browning of armeria due to a Phytophthora crown rot.

Artemisia

Asteraceae

Artemisia, or wormwood, contains species grown for their aromatic, often gray foliage. Some species are entirely herbaceous, while some become woody at the base. *A. lactiflora*, which has dark green leaves, is grown for its flowers and performs well when grown in sunny sites in heavy soil that holds moisture. All other species should be grown with little fertilizer and in dry, sunny well-drained sites; they will tolerate poor soils. *Artemisia stelleriana* is beach wormwood, tolerant of salt spray but needing excellent drainage.

One of the most popular garden varieties, *A. schmidtiana nana* 'Silver Mound,' develops an open center in hot humid summers, causing it to lose its handsome mounding habit. It does best in more northern U.S. climates. Even in the north, however, if watered from overhead, stems at the center of a 'Silver Mound' plant may turn darker gray, wilt and die. This is typically due to attack by either the fungus *Rhizoctonia solani* or the fungus-like *Pythium* spp., which cause the center of the mound to collapse. Affected plants sometimes die out in their centers, leaving a donut-shaped mound. Adjustments in irrigation practices can reduce these problems. *A. schmidtiana* 'Silverado' is less prone to melt-out.

No other diseases are reported for the ornamental species of *Artemisia*, but other members of the genus in the United States are known to be susceptible to various rusts and leaf spot fungi, as well as powdery mildew and downy mildew.

The withering of some of the branches of this artemisia is a symptom of Pythium root rot, which has restricted the water supply to the foliage.

Excessive watering from overhead has led to crown rot in these artemisias.

Rhizoctonia crown rot has killed the center of this clump of artemisia.

Aruncus

Rosaceae

Aruncus dioicus, or goat's beard, grows well in moist environments, in rich soil. The flowers are slightly different on the male and female plants, with the female plants staying attractive longer than the males after flowering. Full sun locations work fine in the northern United States, but afternoon shade and moist conditions are desirable in the South. Wherever it is grown, marginal leaf scorch will result if moisture is not regularly available.

Aruncus is not particularly prone to diseases. Fungal leaf spots caused by *Cercospora*, *Cylindrosporium filipendulae*, *Leptosphaeria aruncii*, *Phyllosticta*, and *Ramularia ulmariae* have been reported in the United States, and Cristulariella leaf spot occurs in Canada. At times powdery mildew may be found on the foliage, causing white colonies mainly on the upper leaf surface. Generally aruncus should be a vigorous grower and trouble-free except for attack by fungi such as *Rhizoctonia solani* (at the stem base) and *Botrytis cinerea* that commonly attack a wide range of plant species. If leaf spots arise, remove the affected leaves and pay attention to how long the leaves sit wet after watering. Watering early in the day to promote prompt drying of the foliage is best if trickle irrigation is not available.

An anthracnose fungus has caused the leaf spotting and blight on this aruncus foliage.

Botrytis canker at the base of an aruncus plant.

Fine white stippling and scorching on aruncus due to spider mite feeding.

Asarum

Aristolochiaceae

There are some deciduous and some evergreen *Asarum* species (wildgingers) from both North America and Europe that are grown for their showy foliage. These are woodland plants and like shady sites with rich, slightly acid, well-drained soils. Moisture must be available at all times. The European wildginger, *A. europaeum*, does best in the northern United States because it is not very tolerant of heat. *A. shuttleworthii* is native to the Southeast; although it lacks winter hardiness, it has heat tolerance.

Wildginger can be affected by leaf galls caused by the fungus *Synchytrium asari*. Fungal leaf spots may be caused by *Ascochyta, Laestadia* or *Plagiostoma* species. Although not yet noted on European wildginger in the United States, the rust fungus *Puccinia asarina* is likely to be found in the western Untied States on native species of *Asarum*. Rhizomes may rot due to infection with *Sclerotinia sclerotiorum*, recognized by the masses of white mycelium on underground stems of ailing wildginger plants. Poorly drained soils may lead to Pythium root rot as well as black root rot caused by *Thielaviopsis basicola*. Given the right environment, wildginger performs well and has no disease problems. Pick off the occasional spotted leaves and, if leaf spots threaten to become a significant problem, adjust watering practices to keep the foliage from sitting wet for long periods of time.

Stunted, blackened root tips on European wildginger due to Thielaviopsis root rot.

Foliar nematode (*Aphelenchoides* species) can attack leaves of wildginger, causing discolored areas that are limited by leaf veins. Affected leaves can be picked off, and the problem can be reduced by preventing the splashing of soil particles onto leaves.

Impatiens necrotic spot virus (INSV) was found in this wildginger showing flecking and distortion at the leaf margins.

Chlorotic splotches and dark line patterns indicate a virus infection on a wildginger leaf.

Discolored areas in wildginger leaves due to foliar nematode (*Aphelenchoides*) infestation (above and right).

Asclepias

Asclepidaceae

Asclepias, or milkweed, includes both weed and ornamental species. *Asclepias tuberosa* or butterfly weed is the native milkweed most often used in perennial gardens. It is well suited to hot, sunny sites. It does well in soils over a wide pH range, but requires good drainage. Aphids are particularly attracted to butterfly weed, as well as other milkweeds. This native plant does not transplant well from the field, so purchase containers rather than trying to move wild plants. Spring emergence is delayed, so mark its position in the garden.

Butterfly weed can be affected by leaf-spotting fungi, but these do not usually cause significant problems. Three species of *Cercospora* have been noted on *Asclepias* spp., with *Cercospora clavata* being the most widely distributed. *Phyllosticta tuberosa* is another fungal leaf spot pathogen of milkweed, and *Phoma asclepiadea* has been found causing stem blight. Two species of powdery mildew have also been reported on butterfly weed.

Asclepias species also are affected by the bacteria *Pseudomonas cichorii* or *Xanthomonas* species, especially during production. Reducing exposure to overhead irrigation, watering early in the day and alternating bactericides (such as copper and a biocontrol product based on *Bacillus subtilis*) may be helpful in reducing bacterial leaf spot on this crop.

Bacterial leaf spots (both *Pseudomonas* and *Xanthomonas* spp. isolated) on asclepias.

Asclepias spp. can also develop rust. One rust disease is caused by the fungus *Uromyces asclepiadis*, which produces urediniospores that can reinfect butterfly weed, and also teliospores for overwintering on this host. In North and South Dakota, another rust (*Puccinia bartholomaei*) has been reported that has two spore stages on butterfly weed and the rest of its life cycle on the marsh grass *Spartina*. Scout for the signs of rust, and initiate fungicide treatments in the nursery if the disease appears. In landscape settings, remove rust-infected leaves as soon as they are detected, and reduce the frequency and duration of overhead watering if possible.

The fungi *Sclerotium rolfsii* and *Rhizoctonia solani* may both cause lower stem rot. Infected plants should be removed and discarded. In addition, do not reuse containers from dead plants unless they have been cleaned and disinfested first.

The oleander aphid is a common visitor to asclepias plantings.

Two different powdery mildew fungi on asclepias, one on the upper and one on the lower leaf surface.

Aster

Asteraceae

The most popular garden forms of *Aster* are hybrids of American natives improved by plant breeders in Europe. Most flower in the summer or fall and deadheading can extend the flowering period significantly. Over time bare patches will develop at the center of aster clumps, so most types should be divided every few years. Most asters need full sun to reduce legginess and to reduce the impact of foliar diseases; a few (e.g. *Aster divaricatus*) can be shade grown.

The main diseases of asters are Verticillium wilt (*V. albo-atrum*), powdery mildew and rusts. The systemic disease Verticillium wilt is more problematic with New York aster (*Aster novi-belgii*) than with New England aster (*Aster novae-angliae*).

Powdery mildew, which causes a patchy white coating, affects plants of different species and varieties to different degrees. This disease is often found on wild asters in the Western United States. Select varieties to grow that appear attractive in late summer in gardens in your locality. A little powdery mildew or rust can be tolerated, as infected plants will still flower year after year. Heavily diseased plants lose much of their ornamental value, however, so fungicides may be used in the latter half of the growing season if highly susceptible varieties are being grown.

Rusts caused by *Puccinia* species cause chocolate brown spore-containing pustules to form on the underside of the leaves, while the upper surface may show yellow spotting. The rust *Coleosporium asterum*, in contrast, exhibits orange-yellow spores. On some varieties, rust diseases may be serious enough to justify preventive fungicide application in the garden. Fungicide applications in nursery and garden center to discourage rust and powdery mildew may be important for asters intended for fall sales. For cultural management of powdery mildew and rust, in addition to variety selection, be sure to space plants well and, if necessary, prune out some stems mid-season to allow good air circulation.

Since asters are such a large group of plants, there are many possible minor diseases. Leaf spotting fungi that may interact with asters include species of *Alternaria*, *Ascochyta*, *Botrytis*, *Cercospora*, *Cercosporella*, *Leptothyrium*, *Ovularia*, *Phyllachora*, *Phyllosticta*, *Placosphaeria*, *Ramularia*, and *Septoria*. Synchytrium leaf galls are another possibility. A disease called black knot, caused by *Gibberidea heliopsidis*, has been reported on aster from various parts of North America. This disease occurs on plants in the Astereae and Heliantheae; symptoms include black lesions on stems, as well as along the main veins or petioles. Scattered tiny black leaf spots appear on the leaves as well. A downy mildew caused by *Basidiophora entospora* is also seen occasionally. This disease can be recognized by the white patches of downy mildew sporulation that appear on the undersurface of leaves. The downy mildew tends to sporulate along the leaf midvein. Providing sufficient spacing between plants and avoiding over-frequent overhead watering will help to reduce humidity around plants and thus reduce severity of most foliar diseases.

Asters are susceptible to *Tomato spotted wilt virus* (TSWV) and *Impatiens necrotic spot virus* (INSV), which can result in necrotic spots and rings on foliage. *Tobacco ringspot virus* (TRSV) may cause stunting as well as systemic chlorotic blotches on aster leaves. *Tobacco rattle virus* (TRV) also occurs on aster in the United States, and *Aster chlorotic stunt carlavirus* has been reported from Canada on New England aster. Nematodes can affect asters, including root knot nematode (*Meloidogyne* spp.) that feeds on roots in the ground and foliar nematode (*Aphelenchoides* spp.) that feeds above ground. Sclerotinia stem rot has been reported on asters, as has Phymatotrichopsis root rot (the latter in Deep South states primarily). Crown gall caused by *Agrobacterium tumefaciens* may affect asters, resulting in swollen areas on the roots or at the root crown. If crown gall or diseases due to viruses or nematodes are found, remove affected plants from the garden or nursery, in order to reduce the chance of spread.

Rust sporulation may appear on both stems and leaves of asters.

Rusts caused by *Coleosporium* spp. on aster show orange sporulation, primarily on the leaf undersurface.

Brown sporulation on aster leaves is caused by a Puccinia rust.

Powdery mildew has coated the aster foliage at right.

Downy mildew caused by a *Basidiophora* sp. shows whitish sporulation along the midvein of infected aster leaves.

Downy mildew causes chlorotic patches to appear on the upper leaf surface of aster.

This swollen mass of tissue at the soil line is the symptom of crown gall disease on aster.

Closeup of an aster root showing browning and decay of the cortex by *Phytophthora*.

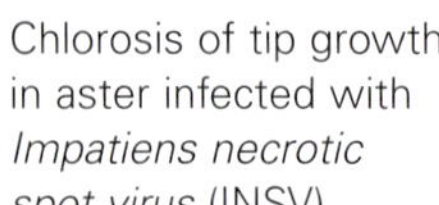

Chlorosis of tip growth in aster infected with *Impatiens necrotic spot virus* (INSV).

Astilbe

Saxifragaceae

Astilbe, or false spirea, includes some of the mainstays of shady perennial gardens. There are many astilbe species and numerous hybrids including the popular *Astilbe* x *arendsii* hybrids from Germany. These contribute delicate spikes of white, pink or red to the garden.

Failure to flower signals a need to divide and replant—astilbes will not flower if they are overcrowded. They are relatively disease-free, but will show symptoms that can be confused with contagious disease if they are grown in dry soils or in bright, windy sites. Leaf edges will develop a reddish brown scorch in midsummer if astilbes are not grown in the moist, shady conditions that they prefer. *Astilbe chinensis* varieties are relatively drought tolerant. Another common stress on astilbe is root injury from the feeding of black vine weevil larvae or Oriental beetle grubs. This can result in stunted plants with chlorotic foliage that are more susceptible to winter kill. Root feeding by black vine weevil can be extensive on astilbe, but the foliage does not show the characteristic notching that is often found on rhododendrons and yews.

Various *Cercospora* species cause minor leaf spots on astilbe around the world. Powdery mildew may rarely occur. Improve air movement around plants if powdery mildew begins to develop. Foliar nematodes may invade leaves through the stomates and feed within the leaf tissue, resulting in vein-limited brown patches that most often are formed on the lower leaves. If foliar nematode injury is observed, remove the leaves showing the injury, thin out dense clumps to improve air drainage, and schedule irrigation to reduce the length of leaf wetness periods. The root knot nematode, *Meloidogyne*, may feed within the roots, causing galls and possibly stunting the above-ground part of the plant.

Astilbes are also subject to attack by *Rhizoctonia solani*, a fungus that grows at the base of the plant. Infected plants will be stunted or wilted and have brown sunken cankers near the soil line. Remove seriously injured plants and protect the rest by watering in such a way that the soil surface dries between waterings. Do not mulch all the way up to the stem of the plant, as this will create an area of high humidity right at the soil surface that favors attack by this fungus. Wilt in astilbe might also be due to infection by a *Fusarium,* but this has not been widely reported.

If roots appear healthy but plants are chlorotic, they might be suffering from *Tobacco ringspot virus* (TRSV). Plants with virus infection should be discarded, as there is no treatment.

Swellings on astilbe roots due to root knot nematode.

Twisting and spotting of an astilbe petiole infected with *Thielaviopsis basicola*.

Scorching at the tips and edges of leaves is common for astilbes subjected to drought stress.

Leafhopper feeding causes yellowing and scorching at leaf tips.

Astrantia

Apiaceae

Astrantia or masterwort should be grown in moist conditions, partial shade and rich soils. It does best in areas with cool night temperatures, thus it is very popular in English gardens. Its fireworks-like blooms are sometimes used as cut flowers, since they can last for up to two weeks.

The leaves of astrantia show typical vein-limited brown patches when invaded by foliar nematodes (*Aphelenchoides* spp.). Although no fungal diseases have been reported on astrantia in the United States, a few leaf spots and a rust have been found in Europe. A fungus identified as *Leptotrochila astrantiae* has recently been found fruiting on yellow-rimmed leaf spots in New York; a foliar disease caused by this fungus has been reported from Europe. Avoiding excessive overhead irrigation and watering early in the day will help to minimize both foliar nematode and leaf spots on this plant.

These yellow-haloed leaf spots were caused by a fungus, *Leptotrochila astrantiae*, on masterwort.

Brown vein-limited spots on masterwort indicate foliar nematode infestation.

Baptisia

Fabaceae

Baptisia, or false indigo, is in the bean family, so the plants sport interesting seed pods after blooming. These American natives are available in various shades ranging from white to cream to blue. Sunny sites are needed for all but the white forms, which can be grown with some shade. Baptisias can tolerate poor soils and warm climates. They are virtually trouble-free, and very drought tolerant, but appear to be especially attractive to voles. Baptisias do not transplant well, so try to set them into an ideal position in the garden on the first try.

Diseases are uncommon on this plant. The foliage will blacken dramatically in response to frost. Minor leaf spots caused by the fungi *Cercospora*, *Marssonina*, *Septoria*, and *Stagonospora* species may be seen on baptisia. Powdery mildew is fairly common, and a Puccinia rust has been reported from scattered locations in North America.

Begonia

Begoniaceae

Most *Begonia* species are not hardy, but there are a few which can be grown as perennials in warmer regions, such as *Begonia grandis* and *B. sinensis*. These can be grown in semi-shade. A moist garden environment and rich soil will allow them to thrive.

Although powdery mildew is common on many of their greenhouse-grown relatives, it has not been a major problem on perennial begonias in the garden. *Begonia grandis* does, however, share with greenhouse begonia crops a susceptibility to *Xanthomonas axonopodis* pv. *begoniae (Xanthomonas campestris* pv. *begoniae)*, a bacterial pathogen that causes severe blighting of leaves and stems, as well as vulnerability to bacterial soft rot. In nursery or garden, plants that develop these problem should be disposed of, and clean stock obtained.

Begonias are also susceptible to Rhizoctonia stem canker. Avoid mulching right up to the stem, and provide thorough watering as needed, rather than providing frequent light waterings that keep the surface of the soil wet.

Impatiens necrotic spot virus (INSV) kills leaf tissue in spots and rings. Dead spots on leaves may also be the result of foliar nematode infection (*Aphelenchoides* spp.).

Bacterial leaf spot caused by *Xanthomonas axonopodis* pv. *begoniae* on a hardy begonia.

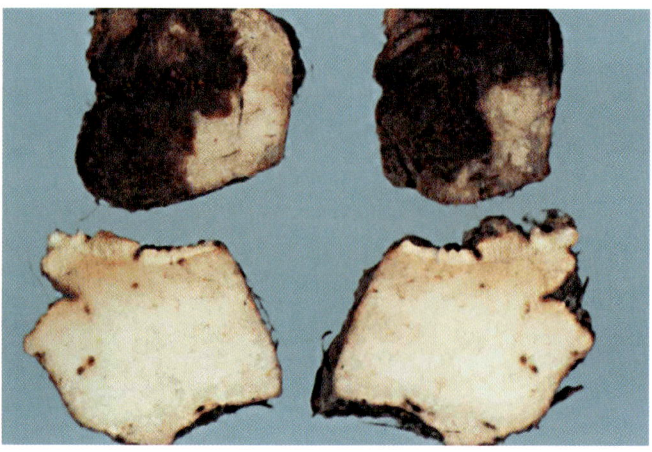

Soft rot of begonia tuber in storage by the bacterium *Dickeya chrysanthemi*.

Belamcanda

Iridaceae

Belamcanda, or blackberry lily, belongs to the iris family. Plants are grown for their showy flowers—even the spent blossoms provide an interesting twist. These plants are prone to many of the same problems as irises. Belamcandas do well in a wide range of garden conditions, but they best appreciate sunny, well-drained sites in rich, sandy loam. Mulching helps them to survive in more northerly climates within their range.

The most common leaf disease on belamcandas is Heterosporium leaf spot, also found on iris. This fungal disease causes oval brown spots that are most disfiguring of foliage in wet springs or in gardens or nurseries with frequent overhead watering. If spotting is light, infected leaves can be physically removed as they appear. Alternaria leaf spot and anthracnose caused by a species of *Colletotrichum* fungus have also been seen on belamcanda. Bacterial soft rot (*Pectobacterium*) can affect plants in hot weather. Fungicides and bactericides are not usually necessary. Gather and dispose of foliage in autumn to reduce the overwinter survival of the pathogens.

Belamcandas have also shown susceptibility to *Tomato spotted wilt virus* (TSWV). If yellow ringspots indicating TSWV appear on the leaves of a newly-acquired plant, it is better to remove it promptly than to leave it in the nursery or garden. A tiny insect called the western flower thrips spreads this virus between different perennials, and many annuals and vegetables are susceptible as well.

Bellis

Asteraceae

Bellis perennis, or English daisy, is considered a lawn weed in some areas and an herbaceous perennial in others. The double-flowered forms are popular as pot plants and garden perennials. English daisies are suited to cool, moist climates.

Excess moisture on the foliage is detrimental to *Bellis* because it is susceptible to a rust disease caused by *Puccinia lagenophorae*. This same fungus can cause spots showing orange rust pustules on greenhouse cineraria (*Pericallis* ×*hybrida*) or the weed *Senecio vulgaris*, called common groundsel.

Wet foliage can also encourage Cercospora leaf spot, Botrytis blight, or a bacterial leaf spot caused by a *Xanthomonas*. Bellis is susceptible to crown rot due to infection by *Sclerotinia sclerotiorum*. It may also develop sickly-colored foliage due to aster yellows (caused by a leafhopper-vectored phytoplasma).

Bellis does best when not overwatered, as it can develop Pythium root rot if soil drainage is not adequate. *Phymatotrichopsis omnivora* may cause root rot in some of the southernmost states. *Rhizoctonia solani* may attack at the soil line to cause a crown rot. To counteract Rhizoctonia crown rot, mulch carefully to avoid holding moisture up against the stem. Roots of English daisies are also susceptible to the galling caused by the root-knot nematode, *Meloidogyne*. If root knot nematodes become a problem in the garden, dispose of affected plants and replace with species known to have low susceptibility to them.

Yellow spores spilling from rust pustules onto an English daisy leaf.

The stunted and chlorotic English daisy on the right is suffering from Pythium root rot.

Bergenia

Saxifragaceae

Bergenia species are usually referred to by their Latin name, since their common name is "pigsqueak". Bergenias have smooth, ground-hugging leaves on plants that flower quite early in the spring; these range from white to various shades of pink. Shade is helpful in southern gardens, but full sun is acceptable in cooler climates. Bergenias do not do well in poorly drained soils. The foliage persists through the winter, sometimes taking on a reddish hue, but it is prone to winter damage. At times the flower buds will also be harmed by winter cold. The black vine weevil, *Otiorhynchus sulcatus*, is fond of feeding on bergenia foliage, and will leave behind neat semi-circular holes along the edges of the leaves.

Due in part to their ground-hugging habit, the *Bergenia cordifolia* hybrids are very prone to infestations of foliar nematode (*Aphelenchoides* spp.). These tiny (1/16" long) non-segmented worms live in the soil and crawl up onto plant surfaces following rain or irrigation. They enter leaves through the stomates, causing chlorosis or reddened patches that expand until they have become bounded by major veins. The only controls for this nematode disease are to pick off the infected leaves as they appear and to avoid excessive irrigation, or overhead irrigation that is applied late in the day.

Bergenias are also sometimes afflicted by an anthracnose disease caused by a species of the fungus *Colletotrichum*. This disease causes a gnarling of the young foliage, as well as a peppering of tiny purple spots on the leaves. The foliar distortion follows from the fungus having killed small areas on the leaves so that they cannot expand normally. Most hybrid bergenias in the nursery trade today are not very susceptible to this disease. Check young leaves carefully for signs of purple spotting or distortion at time of purchase. To keep the disease in check, avoid irrigating plants in such a way that the foliage will sit wet for long periods.

The roots of bergenia can be infected by the fungus *Thielaviopsis basicola*, which causes black root rot. Avoid overwatering or high soil pH, greater than pH 6.2, to keep from favoring this pathogen. In the nursery, soil fungicide drenches may be used to combat it.

Bergenia leaf spotting and tattering due to infection by the fungus *Colletotrichum* sp.

Red and brown patches in bergenia leaves caused by foliar nematode feeding.

Brunnera

Boraginaceae

Brunnera, or Siberian bugloss, is a genus of clump-forming ground covers that show blue flowers all spring long. *Brunnera macrophylla* or false forget-me-not (syn. *Anchusa myosotidiflora*) is a species that does especially well in garden areas that are shady and moist. Its flowers resemble those of *Myosotis*, the true forget-me-not, which can be grown under similar conditions. A number of variegated varieties are also available.

Siberian bugloss is not prone to diseases, although it is attractive to aphids. Most symptoms noted on brunnera are due to the gardener's failure to supply the right growing environment: browning along the leaf edges indicates that the plant has been subjected to overly bright conditions, and lack of moisture will also lead to unhealthy looking plants. The variety known as 'Langtrees' or 'Aluminum Spot' is relatively drought tolerant. Excessive watering during production in nursery containers may lead to root diseases caused by *Pythium* or *Phytophthora* species.

Occasionally brunnera may develop downy mildew (*Peronospora myosotidis*). Downy mildew causes yellow to brown patches in the leaves, as viewed from the top. Turning the leaves over will reveal small patches of downy mildew sporulation that are most easily viewed with the aid of a hand lens. Brunnera can also develop discolored patches in the leaves due to foliar nematode (*Aphelenchoides*) infestation. Check incoming plants for discolored areas in the foliage that might indicate one of these problems and reject suspicious plant material. Both downy mildew and foliar nematode can spread to other plants in the garden by splashing during rain or irrigation. Picking off infested leaves or removing plants and the soil in the immediate area are the only control measures for foliar nematode available in a garden setting.

Leaf scorch and plant stunting on this brunnera is due to a Phytophthora crown rot.

The rounded swellings on these brunnera roots are due to root knot nematode.

The dark, vein-limited patches in this brunnera leaf are caused by foliar nematode.

Caladium

Araceae

Caladium, or elephant ear, is a very tender plant that may be classed as a tropical perennial. The extremely decorative leaves of plants in this genus, which come in a wide diversity of colors and patterns, make caladiums popular in northern gardens in spite of their vulnerability to frost. Plants must be lifted and stored over winter in colder climates. Caladiums require an ample supply of water, and many varieties will do best in partial to full shade.

Foliage problems on caladium include occasional fungal leaf spots due to *Cercospora*, *Colletotrichum*, *Helminthosporium* and *Phyllosticta* species. *Botrytis* and *Botryotinia* spp. may also rot caladiums. Watering early in the day and spacing plants to allow good air movement help to minimize leaf spot impact.

Caladiums are prone to Pythium root rot, which leads to stunting and collapse of the plants; roots may initially appear water-soaked. Several species of water molds may be involved, including *P. irregulare*, *P. myriotylum*, and *P. splendens*. The related genus *Phytophthora* may also cause a soft rot of roots. Corm rot may be caused by *Rhizopus stolonifer*, the common bread mold, or by soft rot bacteria under hot, wet conditions. All of these root problems are associated with overwatering.

Chalk rot, one of the most difficult to control diseases, is caused by *Fusarium* species. This dry rot causes outer portions of the corm to slough off. The fungus *Rhizoctonia solani* can attack the plants at the soil level. Wilting or stunting may also be due to Southern blight: this fungus disease is caused by *Sclerotium rolfsii* and is identified by finding masses of fungal mycelium accompanied by tiny mustard-seed-like sclerotia at the base of the plant. Remove wilting plants and the soil around the stem if any of these diseases are found affecting caladiums. In the nursery, fungicides may be used to counteract both water molds and fungi that cause root rot.

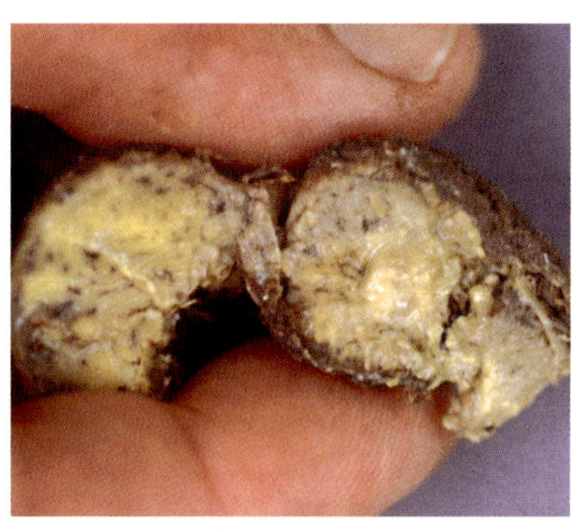

Bacterial soft rot caused by a *Pectobacterium* sp. on caladium.

Pythium root rot on the caladium plant at right has reduced the size and health of the root system.

Fusarium rot on caladium tubers.

Caltha

Ranunculaceae

Caltha, or marsh marigold, includes some plants that are well-behaved in some gardens but irritatingly invasive in moist soils in other regions. The garden calthas are native plants that are happiest in wet areas in full sun. Irrigation and soil rich in organic matter will create a garden environment suitable for these plants. Marsh marigolds provide large buttercup-like flowers (yellow or white) in early spring. These plants closely resemble an invasive non-native, *Ranunculus ficaria*, which also blooms in early spring.

The foliage of caltha may be occasionally marred by leaf spots due to fungi including *Botrytis*, *Botryotinia*, *Cercospora*, *Cylindrosporium*, *Fabraea*, *Phyllosticta*, and *Ramularia* spp. The primitive fungus *Synchytrium* may cause galls on leaves. Other foliar fungal problems commonly occurring on caltha include those caused by a number of rust fungi, such as *Aecidium* and *Puccinia* species. All of these leaf-spotting diseases will be held in check by keeping overhead watering to a minimum, and watering early in the day so that foliage dries before nightfall. There are also a few reports of powdery mildews occurring on caltha in the United States and Canada.

Campanula

Campanulaceae

Campanula, or bellflower, includes garden perennial and biennial species that require good drainage. Alpine types may need sandy humus with some fine limestone added. Some shade at midday improves flowering for campanulas in the Deep South, but they do well in sunny northern gardens. The plants will self-seed and spread readily. Mulching with 2-3 inches of leaves is good winter protection for campanulas in New York and farther north; a lighter covering of evergreen boughs, leaves or straw will suffice in the South.

Campanulas occasionally develop leaf spots caused by the fungi *Phyllosticta, Ascochyta, Ramularia, Septoria* or *Cercoseptoria* spp., or show white spots of powdery mildew. Some rust fungi occur on campanulas, including a *Coleosporium* species (with pine as an alternate host), as well as *Aecidium* and *Puccinia* spp. They are also very susceptible to *Botrytis cinerea*, especially during greenhouse propagation. Minimizing relative humidity and watering carefully to shorten the periods of leaf wetness in the greenhouse are critical for campanula growers. Campanulas may also develop shoot proliferation due to infection by the bacterium *Rhodococcus fascians* (=*Corynebacterium fascians*). If fed on by infectious aster leafhoppers, campanulas can acquire the aster yellows phytoplasma and show distorted flowers and stunted growth. The foliar nematode *Aphelenchoides* may also affect campanulas, causing first yellow, then brown, patches in the foliage. Campanulas showing shoot proliferation or symptoms of aster yellows or foliar nematode should be rogued out of gardens or nurseries.

Campanulas are only rarely seen with virus symptoms, but they are susceptible to *Impatiens necrotic spot virus* (INSV) and have shown chlorosis associated with an unidentified nepovirus during nursery production. Yellow rings and line patterns indicating virus infection are also sometimes observed in nurseries. Distortion of young leaves at the growing tips may be an indication of cyclamen mite infestation; this injury could easily be misidentified as a virus symptom.

Roots are susceptible to *Fusarium* sp., *Phytophthora* sp., and *Rhizoctonia solani*, which may cause root rot. Losses to Rhizoctonia crown rot often are severe when field grown stock is transplanted to containers and irrigated frequently from overhead. Stem bases can be invaded by *Sclerotinia sclerotiorum*, which causes cankers that show white tufts of mycelium of the fungus combined with black knots of fungal matter called sclerotia. Sudden wilt of plants or gradually stunted growth may indicate infection by *Verticillium albo-atrum*, which causes a vascular wilt. Campanulas may develop crown rot if infected with *Sclerotium rolfsii*. These plants also may develop root galls when infested by *Meloidogyne*, the root knot nematode. Root and stem diseases are most common during production, so the use of new pots, fresh potting media and healthy planting stock is critical.

Fungal leaf spot on campanula caused by *Phoma* sp.

Yellow rings and line patterns on campanula indicate a virus infection.

Symptoms of *Impatiens necrotic spot virus* on campanula.

Chlorosis of campanula attributed to a nepovirus.

Stunting and death of the campanula at left was caused by a Rhizoctonia crown rot.

Distortion due to cyclamen mite infestation of a campanula flower head.

Stunting and death of campanulas infected with a *Phytophthora* sp.

Canna

Cannaceae

Canna, or canna lily, includes hybrids at the center of the renewed craze for tropical plants in non-tropical climates. Cannas are not dependably hardy, but should be treated like dahlias, lifting the rhizomes and storing them in moist peat-moss over the winter, in a cool area safe from frost. These plants are sun-loving and appreciate high fertility, so they will perform best in rich, well-drained but moist soils.

Cannas are susceptible to several fungal leaf diseases, including an Alternaria leaf spot and a rust, caused by *Puccinia cannae,* that produces tiny yellow-orange rust pustules accompanied by overall yellowing. A bacterial bud rot caused by *Xanthomonas campestris* pv. *cannae* progresses in the bud before leaves unfurl, resulting in blackening and death of young shoots and leaves. Lesions do not expand in older leaves, but leaves will remain disfigured by infections that occurred within the bud. Decay may affect flower buds, or extend into the stem to cause stem rot. Discard affected rhizomes; do not propagate from material with this disease.

The rhizomes of cannas are prone to Fusarium, Botrytis, and Pythium rots, and *Sclerotium rolfsii* or *Rhizoctonia* may infect the crowns. In the case of *Sclerotium rolfsii*, a cottony fungal growth and mustard-seed-like sclerotia may be observed. Plants affected by the other organisms may appear stunted or wilt without showing any obvious signs of the pathogen. *Radopholus similis*, the burrowing nematode, has also been reported from canna.

Virus diseases on cannas are widely distributed in the horticultural trade. Three potyvirus diseases—canna yellow streak (CaYSV), bean yellow mosaic (BYMV) and canna mosaic (CaMV)—cause mottling and mosaic on the foliage. The badnavirus *Canna yellow mottle virus* (CaYMV) causes yellowing and browning along veins and mottling of leaves. CaYMV is spread through the propagation and sale of infected plants and no insect vector is known. *Cucumber mosaic virus* (CMV) may affect cannas as well. Virus-infected plants may be stunted and perform poorly. Examine the foliage of cannas before purchase for yellow mottling, brown streaks or mosaic that indicate virus infection. Propagators should test their stock and retain only virus-free rhizomes.

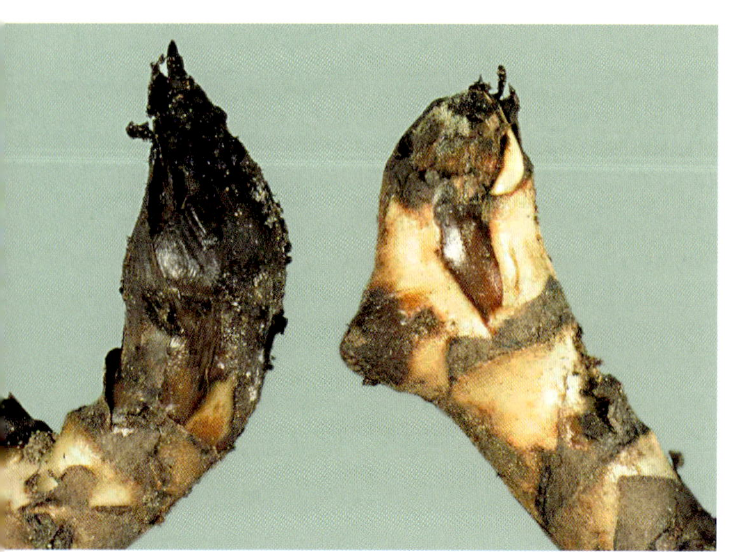

Storage rot of canna rhizomes due to *Botrytis cinerea.*

Internal discoloration of canna rhizomes with Botrytis infection.

Symptoms indicating *Canna yellow mottle virus* on canna.

This curious pattern of holes was made by an insect larva feeding on the canna leaf when it was in bud.

Rust pustules bearing bright yellow spores on canna leaves.

Angular (vein-limited) bacterial leafspots caused by *Xanthomonas* sp. on canna.

Mixed infection of *Canna mosaic virus* and *Canna yellow mottle virus* on canna, causing severe spotting.

Centaurea

Asteraceae

Centaurea, commonly called knapweed or cornflower, includes some weeds as well as some garden plants that will contribute blue flowers accompanied by green or gold foliage in sunny areas with well-drained soils. The commonly grown *C. montana* is an aggressive spreader in northern gardens.

The most common disease on centaurea is powdery mildew. Downy mildew on centaurea is due to the sunflower downy mildew (*Plasmopara halstedii)*, the downy mildew of lettuce (*Bremia lactucae)*, or *Bremia centaureae*. The related pathogen *Albugo tragopogonis* causes white rust (white blister) on *Centaurea* spp., as well as on *Helianthus*, *Gerbera*, *Senecio,* and *Antennaria* spp. and weeds such as *Ambrosia* (ragweed) and *Xanthium* (cocklebur) species. White rust is recognized by white, rust-like blisters that appear on leaves (usually on the undersides) or stems.

Rust diseases caused by numerous *Puccinia* species are found on this genus. Centaurea is also susceptible to leaf spots caused by various fungi including several *Septoria* species. For all of these problems, remove severely affected plants, provide good air drainage, and irrigate with care to keep foliage from sitting wet for long periods.

Centaurea is susceptible to Phytophthora root and crown rot (*P. cactorum*), as well as Phymatotrichopsis and Pythium root rots. *Sclerotium rolfsii* and *Rhizoctonia solani* may attack at the root crown, and *Sclerotinia sclerotiorum* may cause a stem rot. Minimizing excessive irrigation is helpful for avoiding root problems in nursery and garden settings. Centaurea is subject to a vascular wilt caused by the fungus *Verticillium albo-atrum*, which may be acquired from the soil. The other possible vascular wilts, caused by *Fusarium oxysporum* or *F. oxysporum* f. sp. *callistephi*, are more host-specific and would be more likely obtained as latent infections within the plant. Centaureas diseased with one of these vascular wilt pathogens should be removed from the garden as soon as they are detected, and the spot should be replanted with species not susceptible to the pathogen. Centaureas, like other members of the aster family, can also be infected with another systemic disease called aster yellows. Aster yellows causes chlorosis of foliage and virescence of flowers due to the effects of a bacterium-like organism called a phytoplasma living as a parasite in the phloem of infected plants. The phytoplasma disease is spread by the aster leafhopper, which may move the pathogen over from perennial weeds that carry the causal agent. Plants with symptoms of aster yellows should be removed and destroyed.

Powdery mildew on centaurea.

Centranthus

Valerianaceae

Centranthus, or red valerian (also Jupiter's beard), produces flowers that are available in red, white and pink forms. *Centranthus ruber* likes a high pH and a bright sunny location, but does well on soils that are very infertile. These plants will self-sow abundantly and can be very aggressive in gardens in cooler regions of the United States. Deadheading will keep them flowering longer.

Centranthus ruber has few diseases. A Septoria leaf spot appears occasionally (check the center of spots for the tiny dots that represent the pycnidia of the fungus). Phyllosticta leaf spot, also caused by a fungus, appears very similar to Septoria leaf spot. Rhizoctonia root and stem rot can affect centranthus as well. Keep overhead watering to a minimum to avoid encouraging leaf diseases, and do not mulch up against the stem to avoid Rhizoctonia stem rot.

These sunken spots on centranthus are due to infection by *Phyllosticta.*

Ceratostigma

Plumbaginaceae

Ceratostigma, or leadwort, can contribute deep blue flowers and (in *C. plumbaginoides*) maroon fall foliage color to the garden. These plants grow well in moist, well-drained soils in sunny to partly shaded sites. It is normal for plants to emerge late in the garden in the spring. Cover in winter in the northern United States.

There are no fungal diseases reported for ceratostigma, and it is considered to be quite disease resistant. Foliar nematodes (*Aphelenchoides* spp.) cause patches of yellow, red and brown discoloration in the leaves. Avoid excessive overhead irrigation that keeps the foliage wet for long periods of time. When the plant surfaces are wet, foliar nematodes are able to move up from the soil and into the stomates of the leaves. Splashing rain or irrigation will move them throughout a groundcover planting of ceratostigma. Inspect plants before purchase and bring only plants without suspicious patches of leaf discoloration into the garden, particularly if they are to be used as a groundcover. Unless foliar nematode is inadvertently introduced on this or another plant, ceratostigma will probably not be troubled by diseases in nurseries or gardens.

Red or brown patches limited by veins indicate foliar nematode on ceratostigma.

Chelone

Scrophulariaceae

Chelone, or turtle-head, is a genus of North American natives which grow best in acid soils with a high level of organic matter. They can tolerate full sun, but are best in half-shaded, moist, even swampy areas. Some species, such as *C. glabra* and *C. lyonii*, do best in areas with only moderate summer heat. *C. obliqua* is less cold hardy than these two species.

Chelone is susceptible to several fungal diseases including Septoria leaf spot, powdery mildew, and a rust. Irrigate in late morning to midday in nursery or garden in order to avoid extended periods of leaf wetness that favor leaf spot and rust development. Space plants for good air movement to avoid fostering powdery mildew.

The yellow mottling on the young leaves of this chelone plant is due to an unidentified virus.

Septoria leaf spots on chelone have white centers and purple borders.

The chelone plant at right was grown under more crowded conditions, and shows denser growth of powdery mildew.

Chrysanthemum

Asteraceae

The genus *Chrysanthemum* has undergone extensive taxonomic revision, so that many plants have now been split off into separate genera. Here we will discuss the garden chrysanthemum, *Chrysanthemum* ×*morifolium* (syn. *Dendranthema* ×*grandiflorum*). Chrysanthemum is such a popular and widely used perennial that its diseases are very well known.

Septoria leaf spots, caused by a fungus, appear as oval brown spots on mum leaves, with tiny black pycnidia developing at the center of the spots. A *Cercospora* species may also form leaf spots. The fungus *Didymella ligulicola* (anamorph is *Phoma chrysanthemi*, once called *Ascochyta chrysanthemi)* can form irregular brown spots on leaves, or blight the flowers, in a disease best known as Ascochyta ray blight. The pycnidia of *Phoma chrysanthemi* are sometimes difficult to find in the lesions. One-sided rotting of the flower head is often seen in plants affected by ray blight. Another petal blight disease is caused by *Itersonilia perplexans*.

Necrosis of chrysanthemum flower buds and foliage due to Ascochyta ray blight.

Leaf lesions very similar to those seen in Ascochyta ray blight are formed by the bacterium *Pseudomonas cichorii*. Irregularly shaped brown lesions often form along the edge of leaves on susceptible varieties. This disease occurs in production in midsummer but is somewhat more common in the garden during the late summer and

Alternaria blight on lower petals of chrysanthemum.

Leaf lesion caused by *Phoma chrysanthemi*, the causal agent of Ascochyta ray blight on chrysanthemum.

fall. Chrysanthemum is susceptible to powdery mildew, but this disease is not very common.

Two rust diseases are important on chrysanthemum. The first one is *Puccinia chrysanthemi*, the brown rust of chrysanthemum. It causes yellow spots on the upper surface of the foliage, with chocolate brown rust pustules forming on the undersurface opposite these. Brown rust is very contagious, and can spread from plant to plant with rain or splashing of overhead irrigation. Fungicide applications are needed to halt the progress of this disease. A second rust disease, chrysanthemum white rust, is caused by the fungus *Puccinia horiana*. White rust forms pustules that are white, pinkish or buff-colored on the leaf undersurface. Viewed from the top of the leaf, these two rust diseases look alike. The difference is important, however, as the white rust is under federal quarantine and must be eradicated: any sightings of this problem should be reported immediately to your state department of agriculture.

The foliar nematode (*Aphelenchoides* spp.) forms brown, vein-limited wedges in the lower leaves of chrysanthemums; these infested leaves will eventually turn completely brown and dry and hang down along the stem. Immature leaves at the growing tip may sometimes be distorted by foliar nematodes feeding upon this tissue under wet conditions.

Chrysanthemums are susceptible to several soil-borne diseases including crown gall, which is caused by the bacterium *Agrobacterium tumefaciens*. These bacteria enter wounds and interact with the plant cells in a way that results in the formation of large numbers of swollen cells which form a gall, usually on the roots or at the stem base. Galls may also appear on stems or leaves if the tissue has been wounded. If galls are noted on a potted plant, do not transplant it into the garden. The crown gall bacterium has a wide host range, including many different garden plants.

Mums will wilt when attacked by the fungus *Rhizoctonia solani* or by water molds. *Phytophthora* spp. often attack at the soil line, while *Pythium* spp. attack at the root tips. The water mold *Pythium aphanidermatum*, which is most active at high growing temperatures, is especially harmful to chrysanthemum. Black root rot (*Thielaviopsis basicola*) infection is also possible. Different fungicides may be used to protect against these various root diseases during commercial production of chrysanthemums. Use growing mixes with good drainage capacity, and avoid overfertilization.

Chrysanthemums are subject to a Fusarium wilt caused by a host-specific fungus, *Fusarium oxysporum* f. sp. *chrysanthemi*. The affected plants may be stunted and show symptoms of water deficiency: leaves may yellow and wilt on the infected plants. The Fusarium wilt pathogen can live in the soil for years in southern climates, so it is important to avoid introducing this fungus to a garden or nursery. Liming the soil, using certain fungicides, and avoiding ammonium forms of nitrogen in fertilizer all help to counteract Fusarium wilt, but culture-indexed cuttings and resistant varieties are the most reliable management tools. A bacterial disease caused by *Dickeya chrysanthemi* can also harm mums, causing a bacterial wilt. Both the Fusarium wilt and

Crown gall disease may cause abnormal swellings (galls) on roots, stems or leaves of infected mums.

Brown rust of chrysanthemum *(Puccinia chrysanthemi)* is easily identified by the chocolate brown spore pustules on the leaf undersurface, opposite the yellow spotting seen from above.

Chrysanthemum white rust caused by *Puccinia horiana*, showing yellow leaf spots on the upper leaf surface (left) and whitish spore pustules on the lower surface (right).

bacterial wilt develop symptoms at high temperatures and may be latent if weather is cool. Stem rot can be caused by *Sclerotinia sclerotiorum*: this fungus produces fairly obvious white mycelium at the stem base, and also black sclerotia that resemble mouse droppings, either on or inside the stem. Promptly remove plants that show symptoms of any of these diseases. Remove and replace soil around plants that have collapsed due to Sclerotinia stem rot, in order to eliminate sclerotia that would otherwise carry the pathogen over to the next growing season.

Many viruses can affect chrysanthemum, but these are only rarely seen in the trade or in gardens due to virus-indexing programs employed by specialist propagators to eliminate these pathogens from mother stock. *Tomato spotted wilt virus* (TSWV) and *Impatiens necrotic spot virus* (INSV) can cause a wide range of symptoms on mums, including ringspots, stem discoloration and wilt. The highly contagious chrysanthemum stunt viroid can cause severe plant stunting, brittle stems and early flowering. Flowers may be significantly reduced in size, or show color break. *Chrysanthemum stem necrosis virus* (CSNV) causes TSWV-like symptoms, including dark streaks on stems, wilt of branches and dark or yellow rings or spots on leaves. Some varieties show large yellow blotches called "measles", while leaves are sometimes flecked, spotted or distorted. TSWV, INSV and CNSV are each transmitted by thrips.

Wilt and chlorosis in mums due to *Pythium aphanidermatum* root rot.

Foliar nematode injury to chrysanthemum.

Septoria leaf spot on mum.

Dark bacterial leaf spots caused by *Pseudomonas cichorii* on chrysanthemum.

Bronzed patches in leaves of chrysanthemum with *Tomato spotted wilt virus* (TSWV).

Itersonilia petal blight on a chrysanthemum flower head.

Fusarium wilt of chrysanthemum.

Clematis

Ranunculaceae

Clematis, or leather flower, includes some well-known woody vines. In some climates certain *Clematis* species are considered noxious weeds. This large genus also includes some true herbaceous perennial species, including *C. heracleifolia*, *C. integrifolia* and *C. recta*. A rich, light loam is ideal for clematis; the soil should be well-drained, and the plants need some fertilization annually.

Ascochyta blight on clematis. Infection on the stem will lead to clematis wilt.

Diseases are best known for vining clematis and hybrids. They are, for example, prone to a leaf spot disease caused by the fungus *Phoma* (formerly *Ascochyta*) *clematidina*. Ascochyta leaf spots form on leaves, but when infection occurs on a petiole or grows down the petiole into the stem, the fungus can cause a stem canker that leads to a wilt and dieback of the plant. The disease known as 'clematis wilt' results from this fungal attack on the leaves and stems. Ensure good air movement around clematis stems, and remove leaves showing fungal leaf spots as they are detected. The disease must be prevented starting during propagation since a few leaf spots at this stage can amplify into a costly dieback problem later in production. Spores will be spread during irrigation. Space containerized plants and water early in the day to reduce disease.

Botrytis cinerea will attack flowers, stems and leaves of clematis. The fungi *Cercospora*, *Cylindrosporium*, *Phyllosticta,* and *Septoria* spp. may also cause leaf spots on occasion. Powdery mildew colonies sometimes appear on the foliage and rusts due to *Aecidium* and *Puccinia* species may occur. Phymatotrichopsis and Phytophthora root rots may also bother clematis, the former occurring in the southern United States. Virus problems include *Tomato ringspot virus* (ToRSV). This virus is spread by a few species of nematodes. If these nematodes happen to be present in a garden, replanting a new, healthy plant of the same species will not solve the problem, as it will become diseased in turn. If this happens, switch to plants that are not hosts of ToRSV.

The yellow mottling on these clematis leaves is due to *Tomato ringspot virus*.

A rust fungus sporulating on a swelling that it has caused on a clematis vine.

Phyllosticta leaf spot on clematis.

Fusarium cutting rot on clematis.

Convallaria

Convallariaceae/Liliaceae

Convallaria, or lily-of-the-valley, is an old-fashioned favorite perennial that continues to have devotees. It grows in part-shady, moist, moderately rich soils with high pH, but the flowering is best in sunny sites. It is more invasive in the north than in warmer southern climates. Over time, beds tend to die out, so periodic replanting with fresh clumps is desirable.

Like other plants with a dense habit, lily-of-the-valley is prone to *Sclerotium rolfsii*, a fungus causing the disease known as Southern blight. It can be identified by the presence of round, tan, mustard-seed-sized sclerotia that form on the soil or on the stem base. In North America, lily-of-the-valley can have leaf spots caused by the fungi *Ascochyta, Cercospora, Mycosphaerella*, or *Phyllosticta* species or an anthracnose disease caused by *Colletotrichum* or *Gloeosporium* species.

This plant also is prone to the fungal pathogen *Aureobasidium microstictum* (once called *Kabatiella*

A rust disease caused by a *Puccinia* sp. on lily-of-the-valley in Europe.

Stunting and distortion of the growing tip is apparent on the lily-of-the-valley plant at left, which is infested with a foliar nematode (*Aphelenchoides* sp.).

Reddish lesions from the feeding of the root lesion nematode (*Pratylenchus* sp.) are visible on the taproot of these two lily-of-the-valley plants.

microsticta) that is familiar on daylily as the agent of daylily leaf streak. The disease looks very similar on the two plants. At first, small water-soaked spots develop which later have brown streaks with yellow halos—these spots often form from the leaf tip downward. Infected leaves die prematurely. Rust diseases are more common in Europe on this host, but one rust (*Puccinia sessilis*) has been reported from Canada. Foliar nematode (*Aphelenchoides* species) and root lesion nematode (*Pratylenchus* species) may also cause symptoms on foliage or roots, respectively. Avoid excessive overhead watering so that contagious leaf spots and infestations with foliar nematodes are not encouraged.

Spotting due to anthracnose on lily-of-the-valley.

Spots on lily-of-the-valley caused by an *Ascochyta* sp.

Coreopsis

Asteraceae

Coreopsis, or tickseed, includes common garden plants that should be grown in sunny, well-drained garden spots. Some common species, such as *C. lanceolata* and *C. grandiflora,* have narrow, lance-shaped leaves, whereas *C. verticillata* has threadlike leaves. Varieties of *C. verticillata* last longer than some of the others, and have good drought tolerance. To ensure the best blooming performance, remove spent flowers of coreopsis.

Powdery mildew can coat the leaves of coreopsis, even the thread-like leaves of *C. verticillata*. Downy mildew (*Plasmopara halstedii*) can also affect some coreopsis varieties in the nursery trade, sometimes causing black spots on the upper leaf surface and coating the undersurfaces of leaves with white masses of sporangiophores of the pathogen. Bacterial leaf spot (caused by *Pseudomonas cichorii*) is more damaging, causing irregular brown blotches and spots in leaves in warm, wet environments. Bacterial leaf blighting can be quite common during production in the southern United States. Control with bactericides is often poor, so removing symptomatic plants may be the best approach. Leaves are susceptible to fungal problems also: Cercospora, Septoria and Phyllosticta leaf spots, as well as Botrytis blight. *Cladosporium coreopsidis* causes a leaf disease called scab. Reducing severity of foliar diseases can be accomplished by keeping the leaves as dry as possible and watering when the leaves will dry quickly, if they are sprinkler irrigated.

Aster yellows, a phytoplasma disease spread by the aster leafhopper, can cause coreopsis flowers to develop very peculiarly, with green petals rather than the normal yellow-orange. There may also be abnormal multiplication of flower heads, or elongation or stunting of flower parts. The affected plants are usually smaller and yellower than normal, hence the name of the disease. Control of the disease is largely achieved by managing leafhoppers and eliminating weeds that may be reservoirs of aster yellows in the vicinity of the flower planting. Aster yellows is not seed-borne. This disease occurs in gardens more commonly than in production, but it may trouble cut-flower operations in which weed management is not a priority.

A rust disease on coreopsis has been reported in the southeastern and midwestern United States. It is caused by the fungus *Coleosporium inconspicuum*, which spends part of its life cycle on pine.

Sclerotium rolfsii may cause a crown rot of coreopsis and *Alternaria*, *Rhizoctonia solani* and *Sclerotinia sclerotiorum* may also attack at the stem base. *Phymatotrichopsis omnivora*, *R. solani* and other fungi can cause root rot. Good sanitation practices and mulching carefully to avoid holding moisture right around the stem will help to avoid crown rot problems. Plants can also develop a vascular wilt disease due to infection by *Verticillium albo-atrum*, a soil-borne fungus: plants wilt, turn brown and die. Coreopsis with large populations of root knot nematode feeding on their roots may show nutrient deficiency or stunting above-ground.

The whitish growth of powdery mildew is clearly visible on the surface of broad or narrow leaves of coreopsis.

Multiple small green flower heads have replaced some of the normal flowers (see example at right) on this coreopsis affected by aster yellows.

Stunted, virescent (green) flower heads have been formed on this coreopsis plant infected with aster yellows.

The stunted coreopsis plant with yellowed foliage is infected with the phytoplasmal disease aster yellows.

Coreopsis species are also susceptible to *Beet curly top virus* (BCTV), *Impatiens necrotic spot virus* (INSV) and *Cucumber mosaic virus* (CMV). When coreopsis is infected by CMV, the leaves may show yellow rings or round yellow spots. *Bidens mosaic virus* (BiMV) has been found causing a mosaic on *C. lanceolata* in Brazil. As with all virus problems, treatment is not an option and vegetative propagation will serve to spread the virus. Symptomatic plants should be removed from nursery production or garden sites.

The white coating of downy mildew sporulation is evident on the undersurface of an infected coreopsis leaf.

The dark, spreading lesions on this coreopsis are typical of bacterial leaf spots caused by *Pseudomonas cichorii*.

Dark spotting on the upper surface of a coreopsis leaf infected with downy mildew.

Phyllosticta leaf spot appears as dark purplish spotting on coreopsis.

Leaf mottling on coreopsis caused by *Lettuce mosaic virus*.

Corydalis

Fumariaceae

Corydalis includes garden species that are woodland plants, so they appreciate some shade. Some species also need high organic matter. Good drainage is desirable, and a high pH (6.5 or above) is best.

Corydalis has only a few diseases. A downy mildew (*Peronospora corydalis*) has been reported. A fungal leafspot due to *Septoria corydalis* has been noted. Rust diseases of corydalis are *Puccinia aristidae*, which completes its life cycle on grasses where it occurs in the West and Midwest, and *P. brandegei*, from the West. Corydalis is also susceptible to the root knot nematode, *Meloidogyne*. Check containerized plants for root galls before purchase. If detected in the garden, replace diseased plants with other species less susceptible to root knot.

Crocosmia

Iridaceae

Crocosmia (sometimes called Montbretia) is a genus of sun-loving plants that will generate a lot of blind growth if they are not frequently divided (every two or three years). Avoid planting in wet sites. Corms should be handled like gladioli in the northern United States (above Zone 8).

Crocosmias are not generally troubled by diseases, although as close relatives of gladiolus they occasionally are affected by the same pathogens. They are prone to feeding injury by spider mites under dry conditions. A rust fungus that is under quarantine in the United States because of its threat to gladiolus production, *Uromyces transversalis*, is also known to occur on crocosmia. To date, this pathogen has not been found on crocosmia in the United States, but it has been reported on gladiolus from Florida and California.

Stunted, yellowing crocosmias with Fusarium wilt.

Dahlia

Asteraceae

Dahlia is a genus of familiar, spectacular flowers, long in cultivation, which are available in thousands of varieties, sporting many shapes and colors of blossom. In the North, the roots must be lifted in the fall and stored in moist peat. Dahlias are prone to certain insect and mite problems (notably leafhoppers, aphids and spider mites) and also to some diseases, and they require staking in the garden—but their fans forgive them the extra maintenance. One foliage problem, called leaf scorch, resembles a contagious disease, but is actually due to the potato leafhopper feeding on dahlia foliage. The outer edges of leaves turn yellow, then brown, and foliage is stunted and sickly in appearance. These symptoms are due to the effect of toxins in the saliva of the leafhopper.

One of the most troubling and common dahlia diseases is powdery mildew. There are several different powdery mildew species that can affect dahlias in North America. Under favorable conditions, the leaves and stems may become completely coated with the white mycelium and sporulation of this fungus; high susceptibility to this disease is common in dahlia hybrids. Management requires placing plants in sunny locations, spacing them to allow good air movement, and, if show-quality foliage is demanded, may entail application of fungicides throughout the flowering season. In many areas the disease occurs late in the season and can be tolerated.

White smut, caused by *Entyloma dahliae,* is commonly found on dahlia in many parts of the world. The pathogen is a relative of the rust fungi, and produces resting spores inside the leaf tissue that will carry the disease over from year to year if plant debris is not gathered up and burned or buried in the fall. White smut causes angular pale spots in dahlia leaves that turn brown with time; it will be most serious in areas where there is overhead irrigation hitting the foliage.

Other fungi (*Alternaria* and *Cercospora* spp., for example) may cause minor leaf spots on dahlia from time to time. Foliar nematode, *Aphelenchoides ritzema-bosi,* may cause large yellow to brown dead patches in the leaves. Careful management of irrigation can keep the foliage drier and thus reduce the problem with fungal leaf spots and foliar nematode injury. The flowers of dahlias are susceptible to fungal blights caused by *Botrytis*, *Choanephora* and *Stemphylium* species. Removing spent flowers will help

At left, crown gall on dahlia caused by *Agrobacterium tumefaciens*; at right, bacterial fasciation caused by *Rhodococcus fascians.*

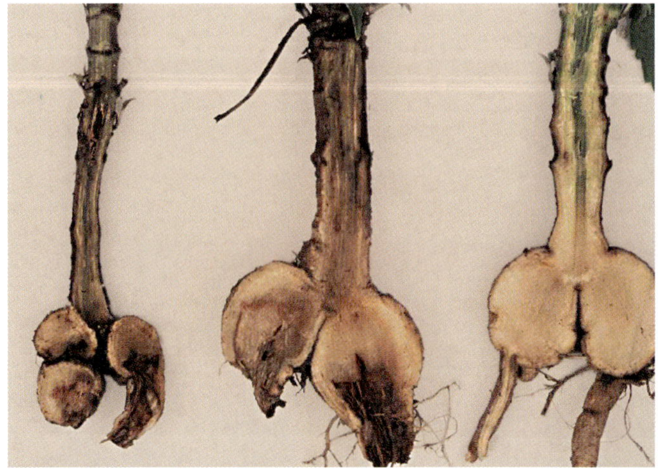

Dahlia tubers showing bacterial soft rot due to *Dickeya chrysanthemi.*

to keep these diseases in check, but also avoid overhead watering to improve flower keeping quality.

Dahlias that wilt may have root rot, or may have been attacked at the soil line by a fungus, or be systemically infected by a fungal or bacterial vascular wilt pathogen. Root rots caused by the fungi *Armillaria mellea*, *Phymatotrichopsis omnivora*, *Macrophomina phaseolina* and *Pythium*, *Fusarium* and *Rhizoctonia* species are all possibilities for dahlia. The first sign of root rot is often grayish, off-color foliage, followed by stunting, wilting and sometimes death. Check planting depth and irrigation volume if root rot develops: deeply planted or over-irrigated plants are the most likely to exhibit root rot.

If attacked at the soil line by *Sclerotinia sclerotiorum*, wilt is more sudden than with the vascular wilt diseases and conspicuous white mold may appear on the surface of the stem base. An additional clue may be found by looking inside the stem, where you may discover black sclerotia (lumps of fungus tissue) that are roughly the size and appearance of rat droppings. Remove infected plants from the garden promptly so that sclerotia are not allowed to fall into the soil and allow overwintering of the pathogen. Wilting dahlias also might possibly be under attack by European corn borer larvae, which feed and later pupate within the stem. Attack at the root crown can also be due to *Sclerotium rolfsii*: look for tiny round mustard-seed-like bodies that identify this fungus.

Dahlias are subject to several vascular wilt diseases caused by the fungi *Fusarium oxysporum*, *Verticillium albo-atrum* or *Ralstonia solanacearum*. In the case of Verticillium wilt, the vascular bundles of the stems or tubers are brown or greenish brown, and the tubers may rot during storage. In the southern United States, slowly wilting plants may be affected by the bacterial wilt disease caused by *Ralstonia solanacearum*. The vascular system in the stem will brown, and show a whitish bacterial ooze when the stem is cut.

Leafy gall on dahlia caused by *Rhodococcus fascians*.

Dahlia leaves showing injury from potato leafhopper feeding.

Black sclerotia and cottony white mycelium of *Sclerotinia sclerotiorum* on dahlia stems.

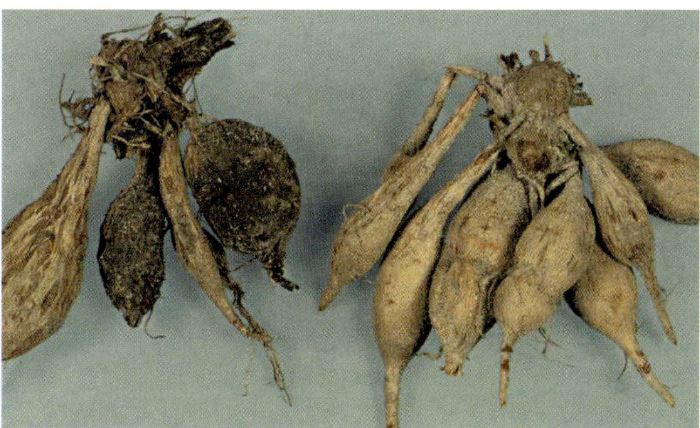

The nematode *Ditylenchus destructor* has caused injury to the dahlia tubers at left.

The root knot nematode, *Meloidogyne hapla*, has caused numerous small galls on the roots of this dahlia plant.

If large lumpy growths are noted on the roots of dahlias, this may be due to crown gall caused by *Agrobacterium tumefaciens*. The thick roots may also be invaded by *Pectobacterium* or *Dickeya* spp., which cause soft rot of tubers or a moist black stem rot. The bacterium *Rhodococcus fascians* may cause leafy galls that look somewhat similar to crown gall. There has even been a report of scab on dahlia tubers, caused by the actinomycete *Streptomyces scabies* that is also responsible for causing a disfiguring disease on potato tubers.

Dahlias are also quite prone to viral diseases. They may be infected by *Dahlia mosaic virus* (DMV), *Tomato spotted wilt virus* (TSWV), *Impatiens necrotic spot virus* (INSV), or *Tobacco ringspot virus* (TRSV). DMV often causes vein banding (yellow discoloration along the veins of the leaves) and stunts plants. Tubers from virus-infected dahlias will carry the disease, so they should not be carried over. DMV may also be carried by seed. Three different types of DMV are known. The *Dahlia mosaic virus* known as DMV-D10 is very widely distributed in the United States: it is integrated into the dahlia genome, and is often latent (causing no symptoms). Efforts to control the disease by aphid management and roguing out diseased plants will thus not be effective until clean stock of dahlias becomes available. *Tobacco ringspot virus* may cause rings in dahlia foliage, light green or yellow in color, on plants that are not especially stunted. TSWV causes yellow ringspots, brown necrotic spots, and flower break in dahlias, among other symptoms. In addition to virus diseases, dahlias may also develop aster yellows, a phytoplasma disease carried by the aster leafhopper, which causes plants to have yellowish foliage and discolored flowers that might be mistaken for viral disease symptoms.

Finally, a number of nematodes have been recorded on dahlia. The root nematode *Trichodorus pachydermus*, the stem and bulb nematode *Ditylenchus destructor*, and the root knot nematode *Meloidogyne* sp. may all affect dahlias, in addition to the foliar nematode *Aphelenchoides ritzema-bosi* mentioned earlier. The roots of plants with poor growth should be examined for indications of nematode feeding.

Brown rot on dahlia tubers, caused by the bacterium *Ralstonia solanacearum*.

Closeup of impatiens necrotic spot (INSV) symptoms on dahlia.

Tomato spotted wilt virus (TSWV) symptoms on dahlia tubers.

Chlorotic ringspots on dahlia leaves with *Tomato spotted wilt virus*.

Stunting and systemic chlorosis of new growth by DMV on dahlia.

Veinal chlorosis caused by *Dahlia mosaic virus* (DMV) on dahlia.

Powdery mildew on dahlia may spot or entirely coat the leaf, or follow veins as shown here.

Dahlia root system showing rot caused by a *Pythium* sp.

Dahlia leaf smut caused by *Entyloma dahliae*.

Dasiphora

Rosaceae

Dasiphora, or shrubby cinquefoil, was until recently in the genus *Potentilla*. *Dasiphora fruticosa* ssp. *floribunda* is the widely grown shrubby cinquefoil previously named *Potentilla fruticosa*. Although a slow growing shrub, this plant is frequently grown with herbaceous perennials. Shrubby cinquefoil is hardy in cold climates and performs poorly south of Zone 7. It is most commonly yellow-flowered, but many varieties with pink, white, or orange-flushed flowers have been produced that bloom from summer through fall. Shrubby cinquefoil should be grown in sunny locations or light shade. Cutting it back in fall will help to maintain a more compact form. It is tolerant of a wide range of soil types, and is drought tolerant, but like many garden plants it performs best in areas with good soil drainage.

Two different powdery mildew fungi may appear on shrubby cinquefoil. The most commonly seen disease on this plant, however, is a rust, *Phragmidium andersonii*, which has also been reported on *Rosa* sp. (rose). A Marssonina leaf spot has also been observed, but fungal leaf spots are not usually a problem on this host. Leaving adequate spacing between plants and avoiding excessive overhead watering are the key cultural strategies for minimizing rust, powdery mildew, downy mildew or other foliar disease problems on shrubby cinquefoil. Allowing the soil to dry between waterings should help to prevent difficulties with root-rotting fungi.

Pale green to tan patches on upper leaf surfaces are a symptom of downy mildew on shrubby cinquefoil.

Rust on shrubby cinquefoil causes tan to white spots surrounded by a red margin on the upper leaf surface.

Delphinium

Ranunculaceae

Delphinium hybrids are quintessential plants for the perennial garden. They are cool weather plants; in order to grow them in the South they must be set out in the fall for a spring bloom. Delphiniums like good drainage and high pH soil, in full sun. They grow well when organic matter and fertilizer are both added to the soil. Remove flower stems after flowering to encourage autumn rebloom. Be careful to transplant to the original soil level, as delphiniums set too deeply are more prone to root rot or crown rot. Delphiniums should be transplanted every 2-3 years to maintain maximum vigor.

Delphiniums are attractive to slugs, and their flower heads are often blackened and stunted by cyclamen mites. Cyclamen mite feeding injury is often mistaken for a disease because the mites, hidden within the buds, are too tiny to see with the naked eye. If cyclamen mites are suspected, break off and remove the affected flower spike from the garden, taking care not to brush it against additional plants. In production of delphiniums, choose miticides that are effective against tarsonemid mites—not all materials that work against spider mites are useful for this purpose.

Severe Pythium root rot on delphinium.

A bacterial leaf spot on delphinium.

Stem base rot on delphinium caused by *Sclerotium rolfsii.*

Yellowing of delphiniums suffering from crown rot due to *S. rolfsii.*

Death of lower leaves and petioles of delphinium due to Botrytis blight.

Powdery mildew on delphinium.

Powdery mildew is one of the most common problems for delphinium. Several different powdery mildew species may affect this plant, causing white patches to appear on foliage and, in extreme cases, leaf yellowing and drop. Gather plant debris in the fall to keep these powdery mildew fungi in check. Botrytis blight can affect the flower buds in wet weather or under overhead irrigation. Leaf spots may be caused by the fungi *Ascochyta*, *Cercospora*, *Diplodina*, *Ovularia*, *Phyllosticta*, *Ramularia*, or *Septoria* spp.; *Synchtrium aureum* causes a leaf gall. Two rusts caused by *Puccinia* species have been reported, and one leaf and stem smut (*Urocystis sorosporioides*). Downy mildew is caused by *Peronospora ficariae*. Leaf spotting due to fungal, water mold or bacterial pathogens can be minimized by watering early in the day to allow foliage to dry before nightfall.

Symptoms may appear on foliage or flowers of delphiniums attacked by *Tomato spotted wilt virus* (TSWV): white ringspots sometimes develop in petals of infected plants. If flowers are green rather than normally colored, delphinium may have aster yellows. This disease, caused by a phytoplasma, may cause abnormal development of flowers or sepals. Aster yellows is spread by leafhoppers. Destroy infected plants and control leafhoppers. *Pythium* and *Rhizoctonia* spp. may attack young seedlings, and several *Pythium* species may attack roots of older plants, especially under conditions of poor drainage. *Phymatotrichopsis omnivora* can attack roots in southern climates. The root crown can be attacked by *Sclerotium rolfsii* or *S. delphinii*, which may show white to dark red or chocolate colored sclerotia the size of mustard seed, or *Sclerotinia sclerotiorum*, which forms larger, black sclerotia. Stem cankers may be due to the fungi *Diaporthe*, *Fusarium*, *Volutella*, or *Phoma*, or to *Phytophthora* species. Stem cankers caused by *Phoma* species are characterized by cracks beginning near the crown; pycnidia form on the discolored areas along the stem. With Diaporthe canker, which is seedborne, lower leaves will turn brown and dry when brown cankers form at the stem base; pycnidia form on the cankers, and mycelium is visible in wet weather. Vascular wilt is due to *Fusarium oxysporum* f. sp. *delphinii*.

Bacterial leaf spot on delphinium caused by *Pseudomonas* sp.

Nematode pests of delphinium include stem and bulb nematode, *Ditylenchus dipsaci*, lesion nematode, *Pratylenchus pratensis*, and root knot nematode, *Meloidogyne* spp.

Wet, black splotches or vein-limited patches in delphinium leaves that sometimes show yellow haloes are due to either *Xanthomonas* or *Pseudomonas* bacteria. The spotted leaves should be removed as they appear. Bactericide or biocontrol sprays may be used during nursery production to inhibit bacterial leaf spots. A bacterial bud and stem rot may also occur, first watersoaking and then blackening the buds or stem, eventually followed by soft rot, leaf yellowing, and stunting of the plant: this disease is caused by *Pectobacterium carotovorum* or *Dickeya chrysanthemi*. The crown gall bacterium, *Agrobacterium tumefaciens*, can cause swellings on roots or at the stem base. Delphiniums with soft rot or crown gall should be disposed of with care, so that the pathogens are not spread to other plants in the garden or nursery.

Faint white ringspots may sometimes be seen on petals of delphinium with *Tomato spotted wilt virus* (TSWV).

On this delphinium the petiole bases have been attacked by *Botrytis cinerea*.

Dianthus

Caryophyllaceae

Dianthus, the genus of flowers known commonly as pinks, is a group of often-fragrant garden denizens. This group includes small plants that are especially popular in rock gardens, as well as the familiar cut flower, the carnation. The non-scented *Dianthus barbatus*, Sweet William, is a biennial that self-sows and does well in the South. Alpine types, of course, do not thrive in southern U.S. heat, but still do well in the northern parts of the United States. There are also many perennial pinks, although these are not very persistently perennial and are likely to need dividing every few years to keep them vibrant. Pinks are sun-loving plants that need good drainage and high-pH soils (slightly alkaline). Diseases of this genus have been well studied because of the importance of the carnation, *D. caryophyllus*, as a commercial florist crop. The pinks grown as herbaceous perennials are not especially disease-prone, so do not be deterred from growing them by the amount of information available about diseases of their florist crop relative.

Flower spot or blight of dianthus may be caused by a number of fungi, including *Stemphylium*, *Botrytis*, *Alternaria* and *Drechslera* species. Both bacteria and fungi may cause leaf spots. The fungal leaf spot due to *Alternaria saponariae* is most often seen on perennials. These spots have a white center and a purple rim; sometimes dark fungal sporulation can be seen on the center of the lesions. Stems may be girdled by the fungus occasionally, resulting in branch rot. Other possible fungal diseases include leaf spots caused by *Septoria*, *Heterosporium*, and *Phyllosticta*. Bacterial leaf spots, caused by *Burkholderia andropogonis*, are sunken oval areas with water-soaked margins. Pick off spotted leaves if possible. The rust fungus *Uromyces caryophyllinus* causes chlorotic spots to form in the leaves, followed by the development of chocolate brown pustules. For either leaf spots or rust, use treatments with appropriate fungicides or bactericides to supplement cultural controls during nursery production. Cultural efforts in gardens and nurseries should focus on keeping the leaves from remaining wet for long periods of time.

Bacterial fasciation due to *Rhodococcus fascians* causes stunting and distortion of plant growth. Galls on roots or at the stem base are usually due to *Agrobacterium tumefaciens*, the causal agent of crown gall disease.

Systemic diseases of dianthus due to viruses or certain bacterial or fungal pathogens are not controllable by chemical treatment. Bacterial wilt caused by *Burkholderia caryophylli* leads to a sudden wilt of individual branches or the whole plant; the roots are also decayed. The affected tissue is sticky to the touch. The bacterium *Dickeya chrysanthemi* can cause a slower wilt. Fusarium wilt is due to the fungus *Fusarium oxysporum* f. sp. *dianthi*. Foliage turns off-color, yellows and wilts. Stems may curl to one side. Internally, brown discoloration is seen in the vascular tissue (xylem). Remove plants if these diseases are

Botrytis flower blight on dianthus.

The straw-colored foliage on this dianthus has developed because of a Fusarium stem rot.

Heterosporium leaf spot on dianthus.

Powdery mildew on the calyx of dianthus.

detected. Do not attempt to replant dianthus in an area where plants have died from Fusarium wilt: the fungus remains in the soil in the form of resistant spores.

Plants that wilt from root rot and lower stem rot may be infected by other species of *Fusarium* besides the one causing vascular wilt. The plants are not as yellowed, and vascular browning is not apparent in the stem beyond the cankered area, when wilt is due to a *Fusarium* other than the Fusarium wilt pathogen. Stem rot may also be caused by the fungus *Armillaria mellea* or by *Rhizoctonia solani*, which infect the roots and the stem base at the ground level. Other root rots may be due to *Pythium* or *Phytophthora* species. *Sclerotium rolfsii* causes Southern blight of dianthus, which appears as a cottony growth at the stem base, accompanied by small round tan sclerotia resembling mustard seeds. To minimize these root and stem problems, keep the mulch away from the base of the stem, and irrigate according to plant needs.

The carmovirus carnation mottle (CarMV) makes plants look pale due to the light yellow mottling of the foliage. Pale streaks may also apear in dark colored petals. *Carnation necrotic fleck closterovirus* (CNFV), a disease sometimes called carnation streak, causes leaves to have reddish or yellow interveinal streaking. Lower leaves may be killed by this virus, which is spread by aphids. Carnation yellows results from dual infection with CarMV and CNFV. Symptoms are more severe when both viruses are present in the same plant. Discard virus infected plants if these symptoms are observed.

Several nematode species have been reported from dianthus, including the root knot nematode, *Meloidogyne*. These can cause symptoms of stunting or nutrient deficiency above ground.

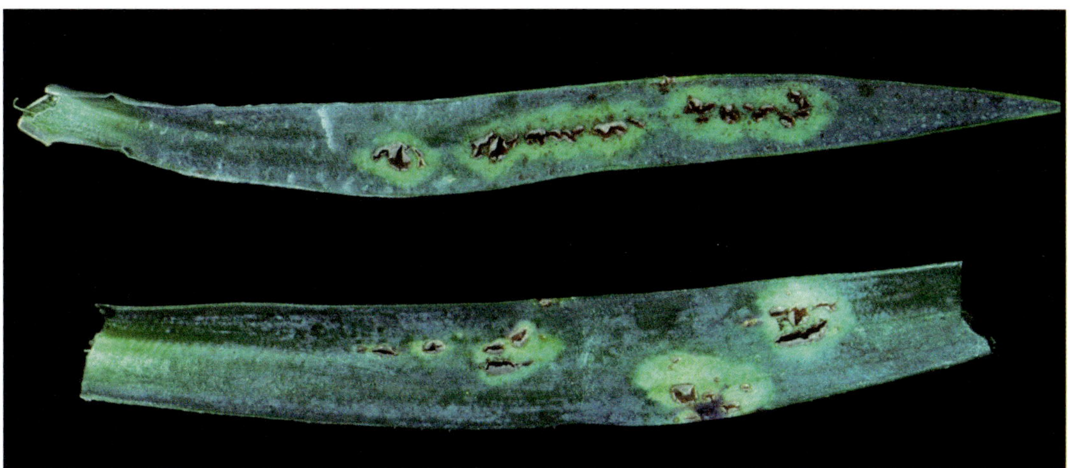

Rust pustules on a dianthus leaf.

Dianthus plants browning from Rhizoctonia crown rot.

Dicentra

Fumariaceae

Dicentra (bleeding heart) is an old-fashioned but ever-popular perennial. In addition to common, fringed and Pacific bleeding heart (*D. spectabilis*, *D. eximia* and *D. formosa*, respectively) there are also other *Dicentra* species with thir own common names, such as *D. cucullaria* (Dutchman's breeches) and *D. canadensis* (squirrel corn). These shade-loving plants thrive in woodland gardens, as they appreciate moist, light soils high in organic matter. If moisture is not in good supply, *D. cucullaria* may exhibit summer dormancy. *D. formosa* has better drought tolerance than *D. eximia*, which is a woodland native from New York to Georgia that does not go dormant in summer. The common bleeding-heart, *D. spectabilis*, grows well in partial shade, and sometimes maintains attractive foliage into the summer, depending on temperature level and moisture availability.

Wilting and yellowing of dicentra due to Verticillium wilt.

Dicentras often show symptoms of *Tobacco rattle virus* (TRV) in nurseries and landscape plantings. Yellow line patterns may appear in leaves as rings, flecks or oak-leaf shapes. Affected plants are likely to be less vigorous than healthy ones, and may not be as long-lived. The virus is spread by certain species of nematodes, and thus may sometimes persist in the soil after an infected plant is removed. Because this virus has a host range that includes a number of other herbaceous perennials (including epimedium), it is important to replant the area of the garden where dicentras have shown TRV symptoms with species that are not known to be hosts of this virus.

Other diseases are seen only rarely. Dicentras are susceptible to Botrytis blight, and may occasionally show leaf spots due to infection by species of the fungi *Cercospora*, *Colletotrichum* or *Stemphylium*. Two downy mildews may attack dicentras: *Peronospora corydalis* and *P. dicentrae*. The diseases produce pale leaf spots that bear fuzzy patches of downy mildew sporulation on the undersurface of the leaf, just opposite the pale spots. In heavy or water-logged soils, Cylindrocarpon, Pythium or Phytophthora root rots occur on dicentras. *Sclerotinia sclerotiorium* or *Sclerotium rolfsii* may cause stem rot diseases. Dicentras are also susceptible to Verticillium wilt and Fusarium wilt. Replanting with resistant species will be important if plants are diagnosed with either of these systemic diseases, as both would leave the soil contaminated with fungus survival structures (microsclerotia and chlamydospores).

Yellow line patterns in dicentra foliage caused by *Tobacco rattle virus* (TRV).

Dictamnus

Rutaceae

Dictamnus, or gas plant, is a garden genus that includes plants that are handsome and long-lived once well established. Its name is descriptive of the plant's unusual talent: it exudes vapors from its flowers and leaves that may be ignited as a party stunt on a calm evening. Gas plants grow well in bright areas with good drainage, as long as nights are cool; they do well only in mountain areas in the South. Dictamnus do not appreciate having their roots disturbed, which makes them difficult to propagate.

Few diseases have been encountered on *Dictamnus* species: they have very reliable garden performance. Leaf spots due to species of the fungi *Phoma* and *Phyllosticta* have been reported.

Digitalis

Scrophulariaceae

Digitalis, or foxglove, includes a number of true perennials, but the garden favorite is the biennial *D. purpurea*, the common foxglove. *Digitalis* species need a steady supply of moisture and do well in light shade in rich soils. Although the foliage of common foxglove may deteriorate after flowering, it need not be defended from diseases throughout the whole summer; removing the plants after seed dispersal is sensible disease management.

A number of diseases have been noted on the common foxglove. Leaf spots may be caused by species of the fungi *Cladosporium* or *Ramularia*, or might be due to an anthracnose disease caused by *Colletotrichum fuscum*. These plants may be attacked at the soil line by *Rhizoctonia solani*, *Sclerotinia sclerotiorum* or *Sclerotium rolfsii*, any of which could lead to wilting and death of plants. Root rot has been attributed to a *Fusarium* species. Be careful not to mulch too closely around the stems of foxgloves to avoid stem and crown rot problems, and do not set transplants too deeply. Foxgloves are also susceptible to Verticillium wilt. If a tell-tale black discoloration is seen in the vascular bundles in the stem of wilting plants, replanting with other species not known to be susceptible to *Verticillium albo-atrum* would be advisable.

One of the most common diseases of foxglove on the West Coast is downy mildew caused by *Peronospora digitalidis*, first reported in 2002. The grayish purple spores of the fungus form on the undersides of leaves directly opposite yellow areas on the leaf top. These areas are often angular in shape and confined between the leaf veins. The disease does not occur every year but is dependent upon wet weather to create ideal conditions for spore formation, spread and infection. Several fungicides are available that will give excellent control of downy mildew on foxglove during commercial production.

Large area of downy mildew sporulation on the undersurface of a foxglove leaf.

Spots caused by downy mildew on a foxglove leaf.

Doronicum

Asteraceae

Doronicum, or leopard's bane, has species with yellow daisy-shaped flowers. *Doronicum orientale* is the most commonly used species in the United States. Warm summer nights and high humidity threaten these otherwise vigorous flowers. They will grow vigorously in a range of soil types in bright or partly shaded areas.

Crescent-marked lily aphid and sawfly may affect doronicum on occasion. It is also possible to encounter powdery mildew. The only other leaf disease reported in North America is due to the foliar nematode, *Aphelenchoides,* which causes yellow or brown patches in the leaves. Root knot nematode (*Meloidogyne*) is also known to affect *Doronicum.* Both of these nematodes may contaminate the soil after an infected doronicum has been removed, so try to replace with plants that are not highly susceptible to foliar nematode or root knot. Reducing overhead irrigation will help to minimize injury from the foliar nematode because it spreads above ground when there is a water film on the plant.

Echinacea

Asteraceae

Echinacea, or coneflower, is a genus of native plants from the eastern and central United States. They are closely related to rudbeckias and were once part of the same genus. These are sunloving plants that can survive hot, dry conditions better than most, and do not require fertilization. The cultivated species will do best in soils at pH 6-7. *Echinacea angustifolia* var. *angustifolia,* blacksamson echinacea, is harmed by standing water but it is more heat and drought tolerant than *E. purpurea,* eastern purple coneflower, which occurs naturally in moist areas. *E. purpurea* tolerates dry conditions once it has become well established. Japanese beetles and thrips may affect echinaceas, as well as the aster leafhopper that vectors the aster yellows phytoplasma.

Coneflowers are susceptible to a bacterial leaf spot caused by *Pseudomonas cichorii* that can be common in warmer parts of the United States where summer rainfall is plentiful. This bacterial pathogen has a very wide host range and rudbeckias and salvia species grown with infected coneflower may also become infected. The fungus *Botrytis cinerea* causes brown leaf spots and stem rot, sometimes with tiny black knots called microsclerotia showing on the dead tissue; this same fungus may also infect flowers, along with *Rhizopus*, which is the common bread mold. Look closely at flowers to distinguish fungal blights from feeding injury caused by the larvae of butterflies and moths.

Alternaria leaf spot causes small dark brown to black lesions that become oblong and develop lighter-colored centers as they enlarge; infections on young plants are often along the midrib. Other fungal leaf spots occur occasionally, caused by *Cercospora*, *Septoria*, and *Ascochyta* species. The white smut caused by *Entyloma compositarum* may also affect *Echinacea* species.

Coneflowers are susceptible to aster yellows, a disease caused by a phytoplasma that is introduced to the plant by the feeding of the aster leafhopper. Infected plants may show bright yellow foliage early in the spring, or reddening of leaves, and sometimes take on a witches' broom appearance when multiple short shoots grow out of the leaf axils. Flower effects are even more dramatic: ray and disk florets may develop green pigment (called virescence), or become distorted into leaf shapes (called phyllody). Keep perennial weeds under control, as these as well as perennial garden plants in the Asteraceae can carry aster yellows over from year to year. Rogue out diseased plants as soon as the symptoms are noticed. *Echinacea angustifolia* seems to be less prone to aster yellows than *E. purpurea*.

Sclerotinia sclerotiorum may cause problems on echinaceas, especially in wet planting sites. When attacked by this fungus, plants will wilt and show dark lesions near the soil line, and blackened, rotting roots. White mycelium as well as irregular-shaped lumpy black sclerotia (similar in size and shape to rat droppings) will be visible on roots and at the crown of dying plants. There can also be direct infection of the flower stems by ascospores in the spring; this will make flowers flop over after the stem decays. *E. purpurea* is more resistant to Sclerotinia crown rot than *E. angustifolia*. The fungus *Sclerotium rolfsii* can also attack the crown of echinacea: it may be recognized by its mustard-seed-like sclerotia. Other root rot problems include several *Fusarium* species, plus *Phymatotrichopsis omnivora* in the Deep South. Wilting plants may have root or stem infection by one of these fungi, or else suffer from Fusarium or Verticillium wilt, which clog the vascular system.

Echinacea is susceptible to *Impatiens necrotic spot virus* (INSV), as well as to other viruses that have not yet been identified.

Flower effects due to aster yellows phytoplasma on purple coneflower. Note the virescent (green) flower parts and poorly developed petals.

Striking contrast between healthy purple coneflowers and virescent ones affected by aster yellows.

Elongated stigmas and sectoring of virescence symptoms in coneflowers with aster yellows.

Angular leaf spots caused by the bacterium *Pseudomonas cichorii* on purple coneflower.

Rhizopus flower blight of purple coneflower.

Septoria leaf spot symptoms on purple coneflower.

Injury to disc florets of purple coneflower by a moth larva.

The yellowing in this purple coneflower foliage was attributed to a misapplication of an herbicide.

Dark brown to purplish vein-limited patches indicate foliar nematode infestation on purple coneflower.

Old foliar nematode injury on purple coneflower.

The blotchy coloring on the leaves of this purple coneflower was caused by *Cucumber mosaic virus* (CMV).

Southern blight caused by *Sclerotium rolfsii* on purple coneflower.

Mottling of purple coneflower leaf due to infection by an unidentified virus.

Echinops

Asteraceae

Echinops, or globethistle, features prickly ornamental plants for sunny, well-drained sites. The commonly grown *Echinops ritro*, southern globethistle, does not require rich soil, and does not need frequent irrigation once established because it is drought tolerant. Flower color is better with cooler nights.

Phytophthora cryptogea root rot may occur on echinops, and also Sclerotinia stem rot and crown rot due to *Sclerotium rolfsii*. Avoid overwatering or mulching too close to the stem. *Cucumber mosaic virus* (CMV) can infect echinops; protect plants from aphids and remove any plants with virus-like symptoms in the foliage.

Epimedium

Berberidaceae

Perennial species of *Epimedium*, or barrenwort, have a knack for covering the ground attractively with heart-shaped leaves and small flowers in various colors, depending upon species. Light shade, soil rich in organic matter, and good drainage ensure successful cultivation; these plants will also grow in heavy shade. The leaves often contribute fall color. The plants may be slow to establish.

Epimedium is a dependable plant not usually afflicted by diseases. Occasional leaf spots on epimedium may be caused by *Alternaria* or anthracnose (*Colletotrichum*) fungi. Epimedium is susceptible to the nematode-vectored *Tobacco rattle virus* (TRV) that causes stunting and ring patterns of yellow or brown in the leaves. Remove infected plants from nursery production. For best success, garden plants should be replaced with non-hosts of this virus, which has a large host range. Epimedium is also susceptible to *Rhizoctonia solani* under high humidity conditions, leading it to develop stem and leaf blight.

The fungus *Rhizoctonia solani* has blighted these epimedium leaves.

The chlorotic rings and line patterns in these epimedium leaves are caused by *Tobacco rattle virus* (TRV).

Eremurus

Liliaceae

Eremurus, or foxtail lily, produces leaves that die down before the plant is in full flower. The flowers can be extremely dramatic in gardens or as cut flowers. Foxtail lily plants need well-drained soils high in organic matter. In very light, sandy soils, drought damage may appear at leaf tips as drying or burning. Mulch plants to protect against frost damage in spring.

Foxtail lilies are bothered by few diseases. The only leaf spot disease reported from North America is caused by the fungus *Myrothecium roridum*. Rust caused by *Puccinia eremuri* has been reported in other countries. Root infections by *Fusarium*, *Pythium* and *Sclerotium rolfsii* have also been found in other countries. Slugs feed on emerging flowers and aphids appear to be drawn to them as well.

Tuber rot caused by a mixed infection of *Pythium* and *Fusarium* on eremurus.

Eremurus tubers with symptoms of crown rot caused by the fungus *Sclerotium rolfsii*.

The leaf tip dieback on this eremurus is due to periods of drought stress.

Eryngium

Apiaceae

Plants of *Eryngium*, or sea holly, grow in dry soils, and tolerate salt, but also do well in average garden conditions as long as the soil is well-drained. For best flower color, eryngiums should be grown in full sun with cool nights. Ideally, transplant sea holly plants while still small, as they develop deep taproots that make them drought tolerant but difficult to transplant.

Slight mottling and chlorosis on stunted plants of rattlensnake master, *Eryngium yuccifolium*, may be due to the aphid-borne *Cucumber mosaic virus* (CMV). Stem rot caused by *Macrophomina phaseoli*, root rot caused by *Phymatotrichopsis omnivora* and a leaf spot due to *Cylindrosporium eryngii* may all affect eryngiums, but these plants are largely trouble-free.

Erysimum

Brassicaceae

Erysimum, or wallflower, has old-fashioned English garden appeal. These plants require neutral to slightly alkaline soil pH and soils with good drainage. Plant wallflowers in full sun except in the southern United States, where partial shade will help them to survive the heat. These are short-lived plants, performing for about 3 years. In the South they can be planted in the fall for spring flowering.

A Xanthomonas leaf spot (*X. campestris* pv. *campestris*) occurs on *Erysimum* spp.; the bacteria can easily be spread from plant to plant on shearing tools, or during irrigation. Related plants such as stock (*Matthiola* spp.) and *Iberis* spp. are also hosts of this pathogen. Plants with this problem should be rogued out quickly from the garden or nursery. Other leaf infections include powdery mildew as well as fungal diseases caused by *Ascochyta*, *Botrytis*, *Heterosporium* and *Pleospora*, downy mildew caused by *Hyaloperonospora parasitica*, and white rust caused by *Albugo candida*. There have been reports of two Puccinia rusts on wallflower. Club root caused by *Plasmodiophora brassicae* causes swollen roots. Root rot in wet sites may be due to *Phytophthora cryptogea* or *P. megasperma*. *Rhizoctonia solani* and *Sclerotinia* species can cause stem rots; to identify *Sclerotinia sclerotiorum*, look for white mold and black sclerotia at the base of the stem.

Black sclerotia of the fungus *Sclerotinia sclerotiorum* on the stem of erysimum.

Bacterial leaf spot due to *Xanthomonas campestris* pv. *campestris* on erysimum.

Eupatorium

Asteraceaee

Eupatorium, known as thoroughwort, boneset or joe pye weed, includes some East Coast natives that are more popular in European gardens than in the United States. These plants perform well in moist, well-drained soils in bright or slightly-shaded areas. They flower late in the summer and into the fall. The most familiar examples are *E. purpureum* (joe pye weed) and *E. coelestinum* (hardy ageratum).

In wet seasons or with overhead irrigation, eupatoriums may have leaf spots caused by the fungi *Ascochyta*, *Cercospora*, *Cristulariella*, *Phyllosticta*, *Ramularia* or *Septoria*, or stem rot caused by *Diaporthe* species. Downy mildew on this host is caused by *Plasmopara halstedii*, the downy mildew that also affects sunflower and many other members of the same family. Rust on eupatoriums may be due to one of a number of species of *Puccinia* and powdery mildew also may affect eupatoriums. Root rot caused by the fungus *Phymatotrichopsis omnivora* or *Macrophomina phaseolicola* (charcoal rot) may occur in hot climates, and *Sclerotium rolfsii* may cause stem rot, making plants wilt.

The earliest recorded plant virus in history may have been on eupatorium. A geminivirus-satellite disease complex reported from *E. makinoi* in Japan is thought to have inspired a poem written in Japan in the year 752. This virus complex is vectored by the whitefly *Bemisia tabaci*. It causes a striking yellow net-vein pattern in affected plants.

Powdery mildew on eupatorium.

Wilting due to a Botrytis canker on eupatorium.

Wilting of eupatorium due to attack at the stem base by *Sclerotium rolfsii*.

Euphorbia

Euphorbiaceae

Euphorbia, or spurge, species are related to the familiar poinsettia used as a holiday potted plant, and, like poinsettia, have showy bracts rather than true flowers. *E. characias*, the Mediterranean spurge, may be invasive in some climates, so seedlings should be controlled. *E. cyparissias* or cypress spurge can also reseed itself or spread by runners. The genus *Euphorbia* contains primarily sun-loving plants that can for the most part tolerate or thrive in light soils, but will not tolerate drought. *E. epithymoides*, the cushion spurge, and *E. wallichii,* the Wallich spurge, need some afternoon shade in southern gardens.

There are only a few diseases reported for *Euphorbia* spp. in the United States. Leaf miner attack might be mistaken for a foliar disease at first glance. Rust due to *Melampsora euphorbiae* is fairly common; a fungal leaf spot caused by a *Cercospora* species may also be seen. The wood spurge, *E. amygdaloides*, is powdery mildew prone in areas of high humidity. Phytophthora root rot may affect euphorbia, and *Phymatotrichopsis omnivora* can also cause root rot in the south. The fungus *Rhizoctonia solani* sometimes causes stem rot, making plants wilt after attack at the soil line. Avoiding diseases in euphorbias is primarily a matter of avoiding excessive irrigation. Excessive foliar wetness will predispose foliage to leaf spots, and excessive soil moisture will predispose plants to Phytophthora root rot. *Poinsettia mosaic virus* (PnMV) has been reported from euphorbia.

Euphorbia wilting from a Phytophthora root rot.

Stem base canker on euphorbia caused by a *Fusarium* sp.

Powdery mildew on a euphorbia.

Mottling in euphorbia due to *Poinsettia mosaic virus*.

Spotting caused by leaf miner tunneling in euphorbia.

Euryops

Asteraceae

Euryops includes subshrubs with bright yellow flowers which are not very cold tolerant and thus can only be used successfully in warm areas. *E. pectinatus* is the species most often grown. Euryops do well in containers in sunny areas, or areas with shade late in the day. They require good drainage.

Only a few problems have been noted on *Euryops* spp. Root diseases caused by fungi include a Fusarium root rot and Armillaria root rot. Powdery mildew has been reported from both coasts.

Cold damage has caused distortion of this euryops flower.

Powdery mildew on euryops foliage.

The white cottony material at nodes on euryops stems is natural, and does not indicate an insect infestation or a fungal problem.

Farfugium

Asteraceae

Leopard plant has natural yellow spots.

Farfugium (previously named *Ligularia*) species are big plants with very yellow flowers or showy leaves to offer the gardener. They do best in moist soils, or bog gardens, in cool climates. Farfugiums are relatively difficult to grow well in the southern United States, where they will need access to soil moisture and some afternoon shade. They should be sheltered from wind. Flowers of *F. dentata* (summer ragwort) and leaves of *F. stenocephala* may wilt and recover in the heat of summer even when moisture is available to the roots.

F. japonicum aureo-maculatum, the leopard plant, comes complete with natural yellow spots that are not a sign of disease. Purple or brown spots, however, do indicate a fungal infection. Most commonly, an anthracnose disease caused by a species of the fungus *Colletotrichum* is seen on farfugium. Cercospora leaf spot has also been reported. Sometimes, small dark brown to black dots can be seen in the browned area. These are structures in which the fungus is forming new spores, which may spread to other leaves and plants. Fungicides can be applied to protect healthy leaves. *Impatiens necrotic spot virus* (INSV) may cause dark ring spots on farfugium leaves. During greenhouse propagation, use yellow sticky cards to monitor for the presence of thrips, which vector INSV, and watch plants for suspicious virus-like symptoms.

Snails and slugs may afflict farfugium. Stem rot and death of plants is caused by *Rhizoctonia solani* and can start as a root rot. Southern blight can also cause stem rot and plant death. The disease is caused by *Sclerotium rolfsii* and usually forms a mass of white fungal growth at the base of the plant during hot, wet weather. The fungus forms spheres, white at first and later changing to dark brown, which allow it to survive in soil and plant debris. These structures are called sclerotia, and they are about the size and shape of mustard seeds. Fungicides can be applied to protect plants, beginning when they are emerging in the spring.

Irregular brown spots caused by the anthracnose fungus *Colletotrichum* sp. on farfugium.

Purple-rimmed leaf spots caused by *Alternaria* sp. on farfugium.

Zonate ringspots caused by *Impatiens necrotic spot virus* (INSV) on farfugium.

Ferns

Phylum Pteridophyta

Ferns are very different from the other plants in this book: they have leaves, but they are not flowering plants. Ferns reproduce using spores rather than seeds. The spores are produced on specialized fronds or on the undersides of leaves. The boldly shaped foliage of many ferns grows in attractive clumps, making them ideal companions for herbaceous perennial flowers, particularly in the shade garden. Several genera, including *Dryopteris* (wood fern), *Athyrium* (lady fern), *Cyrtomium* (holly fern) and *Polystichum* (holly fern, shield fern or sword fern), include native or Asian species that are strong garden performers in temperate climates. In the warmer climates of south Florida and coastal California, additional kinds of ferns are popular, including *Cyathea cooperi*, the Australian tree fern, which withstands light frosts but may be slow to resume growth after a more significant cold period.

Brown to purple discolored patches are caused by foliar nematode infestation on ferns.

Someone unfamiliar with the unique features of ferns might confuse the spore producing areas with disease symptoms or signs, particularly since they usually turn brown at maturity. However, the rows of sori (spore cases) on the undersides of fern fronds are arranged with a neat precision that would not be expected for fungus fruiting bodies, which are produced more randomly. Although some ferns (e.g. *Dryopteris* and *Polystichum* spp.) are especially tough, take care to locate *Athyrium* (including the popular multicolored Japanese painted fern, *A. nipponicum* 'Pictum') in areas away from foot traffic, as the leaves of these and many other ferns are quite brittle. Shade gardens with rich soil are best for many of the commonly cultivated ferns, but some, including *Dryopteris affinis* (golden-scaled wood fern) and *Osmunda cinnamomea* (cinnamon fern), may be used in sites that receive some sun. Well-drained soil is essential for most ferns—rotting of the crowns may occur if plants sit wet over winter.

Distortion (upper photo) and yellow blisters (lower photo) on Christmas fern caused by *Taphrina polystichi*.

The most common disease on ferns is probably web blight due to the fungus *Rhizoctonia solani*, which is especially likely to occur during greenhouse culture or in very humid climates. A fungal leaf spot caused by a *Mycosphaerella* species has been reported on maidenhair fern from Florida. A bacterial leaf spot due to a *Pseudomonas* species has been noted on holly fern during nursery production. Foliar nematodes are a problem on many ferns, including lady fern (*Athyrium* spp.), royal fern (*Osmunda regalis*) and ostrich fern (*Matteuccia* spp.). Pythium root rot may be a problem on these same three ferns.

One curious disease frequently seen on Christmas ferns (*Polystichum acrostichoides*) is a leaf blister caused by *Taphrina polystichi*, a relative of the fungus that causes a disease on peach known as peach leaf curl. If individual plants with leaf blister symptoms are observed in a Christmas fern planting, the affected fronds or affected plants should be immediately removed to curtail spread of the causal fungus. Overhead watering should be avoided, as this will encourage leaf blister development.

Ferns are also susceptible to viruses. Lady ferns with *Cucumber mosais virus* (CMV) infection may be stunted, and the foliage may be malformed. Holly fern foliage may show yellow mottling when the plants are infected by an unknown virus.

Chlorosis on holly fern caused by an unknown virus.

Web blight on fern caused by *Rhizoctonia* species.

Rhizoctonia leaf spots on holly fern.

Bacterial leaf blight caused by a *Pseudomonas* species on holly fern.

Filipendula

Rosaceae

Filipendula, or meadowsweet (also known as false spirea), is native to moist natural areas. Accordingly, all but the European *F. vulgaris* will need a constant supply of moisture, as well as high pH soil (7.0-7.5) in the garden. There are some very showy large filipendulas and also dwarf forms. They grow well in full sun or with some light shade, and benefit from organic soil additives that help to hold water in the soil. Marginal browning of the leaves is a sign of drought in this genus, and the leaves may become entirely necrotic if the stress persists. Some filipendulas are as attractive to Japanese beetles as their rose relatives.

Filipendulas are prone to a few leaf spots caused by species of the fungi *Cylindrosporium* and *Septoria*. The most common disease is a powdery mildew that can cause bruised-looking, reddish areas to appear in the leaves or coat leaf surfaces with whitish growth of the fungus. Occasionally filipendula is affected by the rust *Triphragmium ulmariae*.

White coating of powdery mildew on filipendula.

Virus-like symptoms on filipendula.

Fragaria

Rosaceae

Fragaria, or strawberry, is a very familiar genus because of its much-appreciated fruit-bearing species. New varieties chosen for their ornamental flowers are also becoming popular today. There are important hybrids on the market that are crosses between *Potentilla palustris*, the marsh cinquefoil, and different *Fragaria* (strawberry) species. One example is the variety marketed as Pink Panda®, a cross between *Fragaria chiloensis* and *Potentilla palustris*. Strawberries should be given garden locations in full sun or very light shade, and provided with good soil drainage. Mulch plants to avoid winter injury, which is particularly likely in areas with strong fluctuations in temperature during the winter.

Strawberries grown as ornamentals are susceptible to a wide array of diseases. Good air movement around the plant and proper siting in sunny locations will do a great deal to alleviate foliar diseases. Flowers are susceptible to Botrytis blight, especially in periods of cool, rainy weather. White patches on the leaf surface and upward leaf curl may indicate powdery mildew, which is more severe if plants are not grown in full sun. Several fungal leaf spots may be troublesome. A leaf blight caused by *Dendrophoma obscurans* occurs primarily on older leaves. Large wedges of brown dead areas at leaf tips may be due to leaf blotch caused by *Gnomonia fructicola*; and pale-centered spots with purple borders on upper foliage indicate Mycosphaerella leaf spot, caused by *Mycospherella fragariae*. Phomopsis leaf spot, caused by *Phomopsis obscurans*, is also marked by purple borders. A leaflet blight caused by *Hanesia lythria* may cause individual leaflets to wilt when they are attacked at the petiole. Bacterial leaf spot caused by a *Xanthomonas* species has also been observed.

The crown of the plant is susceptible to anthracnose caused by several species of *Colletotrichum*. Runners may show black lesions and are sometimes killed; crowns may have a reddish-brown internal discoloration. Anthracnose is a summer disease that can be minimized by avoiding overfertilization or extended periods of leaf wetness. All leaf spotting due to fungi or bacteria is reduced by careful watering practices that keep the foliage as dry as possible.

On strawberry roots, infection with the water mold *Phytophthora fragariae* can be a seriously debilitating problem, causing a disease known as "red stele". Longitudinal slices through infected roots will show a reddish discoloration of the xylem that forms the core of the root. The affected plants will be stunted and wilted, and the older foliage will show marginal yellowing or scorch. Plants grown in well-drained soil will not be prone to this problem. Similar symptoms without the red xylem discoloration may be caused by Verticillium wilt. If either of these diseases is encountered, new plants should be obtained and established in a different garden area because the soil will remain contaminated with *Phytophthora* or *Verticillium* spp.

In some cases, several root rotting pathogens, including *Fusarium*, *Pythium* and *Rhizoctonia* species, may come together or act in concert with nematodes to cause a syndrome known as black root rot. The affected plants are stunted and may decline; examining the roots will show an often-extensive black discoloration. *Sclerotinia sclerotiorum* may cause a crown rot: if plants are wilting, look for black sclerotia on the dead stems to identify this problem. Poor

Bacterial leaf spot on strawberry due to a *Xanthomonas* species.

growth of *Fragaria* spp. may also be due to feeding of high populations of nematodes, including species of *Belonolaimus*, *Meloidogyne* and *Pratylenchus*, particularly in sandy soils. Avoid infested areas when replanting strawberries.

Phomopsis leaf spots have a distinct purple border on strawberry leaves.

Sclerotia of *Sclerotinia sclerotiorum* on the stem of a strawberry killed by crown rot.

Gaillardia

Asteraceae

Gaillardia, or blanket flower, is a genus of colorful, daisy-shaped flowers with tolerance for drought. They grow best in full sun, in well-drained soils of moderate fertility, but tolerate even very infertile soil. Deadhead to encourage maximum flower output. Plants will need division every few years to remain robust. The center of the plant may die out naturally, but an outer ring of growth is left to make transplants from.

Gaillardias are subject to a variety of foliar and root diseases. Occasionally they will show mysterious round yellow leaf spots that gradually turn brown—these are a symptom of white smut diseases caused by three *Entyloma* species. Thick-walled resting spores (called ustilospores) that form within the brown spots allow the fungus to survive over winter, so the spotted leaves should be removed from the nursery or garden, once detected. A coating of white spores also forms on the surface of the leaf lesions. White smut may also affect other closely related plants. Generally this disease is a problem under nursery production conditions, in which plants are closely spaced and there is frequent overhead watering. The same fungicides that are effective against rust diseases work well against white smut; systemic materials can be alternated with protectant materials in a management program. White smut is less of a problem on gaillardias once they are in a landscape setting. Other fungi may also affect the foliage of gaillardias. Septoria leaf spot is fairly common, but Phyllosticta and Ramularia leaf spots are reported more rarely. Bacterial leaf spot may be due to infection by a *Pseudomonas* species. Several kinds of rust have been reported on gaillardia in California. Assorted powdery mildew fungi may affect the foliage.

Root and stem base diseases due to *Pythium*, *Sclerotinia*, *Thielaviopsis* or *Rhizoctonia* species may lead to wilting, so avoid overwatering and mulching up against the stem of gaillardias. Along with other members of the Asteraceae, gaillardia is susceptible to the phytoplasma disease aster yellows, which can cause chlorosis of foliage and poor quality flowers. Unusual distortion or patterned discolorations in leaves might be due to virus infection: impatiens necrotic spot (INSV), tomato spotted wilt (TSWV) and cucumber mosaic (CMV) have been reported on gaillardia.

Cercospora leaf spot on gaillardia.

Blighting of gaillardia caused by the bacterium *Pseudomonas viridiflava*.

Early stages of white smut on gaillardia leaves.

White smut caused by *Entyloma calendulae* on gaillardia leaves.

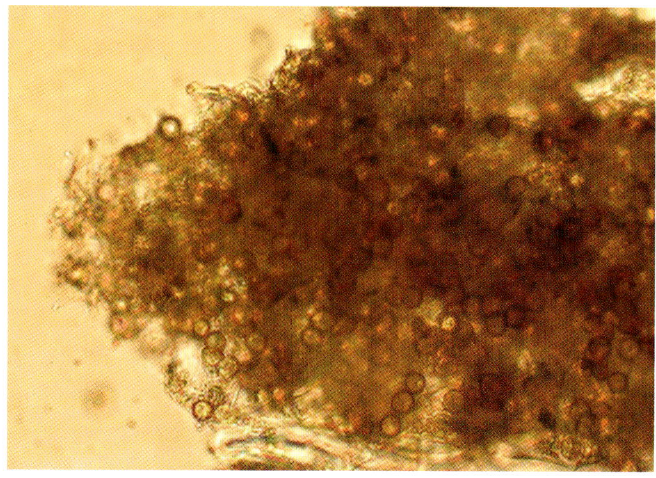

Necrotic gaillardia leaf tissue examined under the microscope will show spores of the white smut fungus.

Root stunting and decay on gaillardia caused by a mixed infection of *Thielaviopsis basicola* and *Rhizoctonia solani.*

Septoria leaf spot on gaillardia.

Galium

Rubiaceae

Galium or sweet woodruff is represented in the garden by one species, *G. odoratum*, which provides a scented, flowering ground cover for humus-rich, slightly acid soils. These plants are aggressive growers and can be invasive in some circumstances: some shade helps to keep them in check in the northern United States, and helps them to tolerate heat in the South. Sweet woodruff will go dormant in summer if the water supply falters; it grows best in moist conditions.

Sweet woodruff does not usually develop disease problems. A few fungal leaf spots, including Septoria, Phyllosticta and Cercospora leaf spots, are fairly common. Downy mildew (especially common during production in the Pacific Northwest), powdery mildew and rust can also affect this plant. One of the most likely diseases to be encountered on sweet woodruff is a foliar blight caused by *Rhizoctonia solani* that causes all or portions of leaves to collapse and turn brown. Irrigate thoroughly and less often, and in late morning or midday, so as to keep the foliage from sitting wet for long periods and thus fostering Rhizoctonia blight. This disease can be a problem in nursery production as well as in the landscape. Keep an eye out for dodder infestations, as this parasitic plant can initially be very inconspicuous in a planting of galium.

Rhizoctonia blight on foliage of sweet woodruff.

Phyllosticta leaf spot on sweet woodruff.

Gaura

Onagraceae

Gaura, or beeblossom, includes mainly North American natives. *G. lindheimeri* is widely grown for its loose spikes of white flowers that develop from pink buds. These have good drought tolerance once they have established deep taproots. Gauras are also tolerant of high temperature coupled with high humidity, which makes them good plants for southern gardens. For best bloom, plant them in well-drained soil high in organic matter, with good moisture supply, and cut back flower stems after blooming. Gauras bloom season-long in the southern United States, but only come into bloom in late summer in the North. Full sun or partial shade is acceptable.

Although not especially disease-prone, *Gaura* is vulnerable to the leafhopper-vectored phytoplasma disease, aster yellows. Spittlebug larvae may feed on the stems within foamy protective masses that give the insect pest its common name. The foliage may be affected by downy mildew, powdery mildew, Botrytis blight and rust. Fungal leaf spots caused by *Cercospora* and *Septoria* species also occur on this host. Quite often, growers are dismayed to note small purple-pigmented spots on gaura in early spring: these are a natural trait of the plant, however, and not indications of a contagious disease, so there is no need for a control action.

Small reddish spots on the leaves are natural to gaura.

White masses of foam on gaura stem produced by spittlebug nymphs.

Geranium

Geraniaceae

Geranium, or cranesbill, is the genus of hardy geraniums, while the florist's geraniums are *Pelargonium* spp. that are found in the same family. Cranesbills are plants for full sun or partial shade. They will need good drainage but grow well in various types of soil. Some shade, plus additional soil moisture, is helpful in southern gardens. Many different species are used in gardens, some of them American natives. Hardy geraniums do best in moist soils, but one species, *G. macrorrhizum*, is a very drought tolerant groundcover. *G. sanguineum* is a heat tolerant, drought resistant plant good for both southern and northern gardens. It is recommended that foliage be cut back after flowering for a number of species. Flowering may taper off during midsummer. Removing flowered stems and old leaves stimulates the plants' productivity.

The hardy geraniums, especially *G. sanguineum*, are prone to bacterial leaf spots that are caused by the same pathogen that causes bacterial blight on the greenhouse crops of the florist's geranium (*Pelargonium* x*hortorum*). Visually, these bacterial leaf spots due to *Xanthomonas campestris* pv. *pelargonii* (*Xcp*) are hard

Alternaria leaf spot on geranium.

Geranium foliage collapse from Pythium root rot.

Phyllosticta leaf spot on geranium.

to distinguish from the spots caused by species of *Alternaria, Botrytis, Phyllosticta, Cercospora, Ramularia* or other fungi. Because of the likelihood of the *Xcp* being carried on perennial *Geranium* species, greenhouse growers of pelargoniums are cautioned to keep the perennial plants well away from the zonal geranium crops.

Geranium carolinianum may show a leaf gall caused by a fungus, *Synchytrium*. The foliar nematode, *Aphelenchoides*, is seen on *Geranium* fairly frequently, causing yellow, red or brown patches between veins. Reducing overhead irrigation frequency and picking off symptomatic leaves will slow the spread of this pest. Pale patches on leaves that are bounded by the major veins may be due to downy mildew, which will produce the "down" of sporangiophores on the underside of the leaf. Downy mildew is seen often on *Geranium maculatum* or *G. carolinianum*. A white coating due to powdery mildew may occur on *Geranium* spp. as well, even though powdery mildew is very rare on their close relatives, the pelargonium hybrids that are grown as the florist's geranium. A number of the rusts affecting *Geranium* spp. are reported only from Hawaii and the western United States, but *G. maculatum* sometimes shows rust infection in the East. Hardy geraniums are susceptible to the vascular wilt caused by *Verticillium dahliae*, but this disease has not been reported in the United States. Root knot nematode, *Meloidogyne*, may form galls on the roots of *Geranium*. *Cucumber mosaic virus* (CMV) has also been detected in *Geranium*.

Xanthomonas leaf spot on geranium.

Powdery mildew on geranium.

Downy mildew of geranium.

Red-rimmed feeding damage caused by plant bugs on geranium may easily be mistaken for fungal leaf spots.

Foliar nematode causes purplish red to brown patches in the leaves of geranium.

Geum

Rosaceae

Geum, or avens, is a genus of rosaceous plants that grow better in the northern United States than the South. A steady supply of moisture, good drainage (particularly in winter) and full sun or some afternoon shade suits them, and they thrive in fertile soil. Cutting back after flowering will stimulate a second flush of blooms. Mulch lightly, to protect against winter injury. Annual division is beneficial for plant vigor.

One of the most frequently seen diseases on geums in nurseries is downy mildew, possibly because frequent overhead irrigation in production areas creates cool, moist conditions that favor the disease in the springtime. The most widespread downy mildew is *Peronospora potentillae*, which can also affect potentilla and a few other plant species.

Downy mildew can fool the gardener since infected plants look as if they have a nutrient deficiency—lower leaves show yellowing or purpling as if they are low in nitrogen. A closer look may show the presence of patches of discoloration that are limited by the main veins. If you turn the leaf over, you may be able to see a fuzzy growth on the leaf undersurface, opposite the discoloration—this is the sporulation of the downy mildew organism. Foliar nematodes may cause similar symptoms in geums, but this fuzz of mildew sporulation will be absent. Picking off and destroying affected leaves is a good practice in either case. Other foliage problems for geum include powdery mildew and some fungal leaf spots caused by *Cercospora, Phyllosticta, Ramularia* or *Septoria* species. Leaf galls may be caused by the fungus *Synchytrium*. A few rust and smut fungi are reported from the United States, but these are not widespread problems.

Aster yellows, a systemic disease due to a phytoplasma that is vectored by the aster leafhopper, can affect geum. The root systems of plants that are doing poorly should be checked for the tiny galls that indicate root-knot nematode infestation. Root systems with conspicuous discoloration might be under attack by a *Phytophthora* species that is favored by poor drainage and deep planting.

Necrotic patches on geum leaves caused by downy mildew.

Vein-limited patches of discoloration caused by foliar nematodes in a geum leaf.

Gladiolus

Iridaceae

Gladiolus hybrids (*G.* x *hortulanus*) are familiar to gardeners. In areas where they are not hardy, corms should be dug up before frost and dried prior to winter storage. Glads are grown in sunny sites, in well-drained soils. Thrips are a common insect problem for this genus. Often their feeding injury (small whitish streaks on the leaves and flowers) is confused with virus symptoms. Look closely for dark dots of the thrips' droppings on the lesions, and notice the overall pattern: virus symptoms are more random, while thrips feeding can be very uniform within a gladiolus planting.

A bacterial leaf blight of gladiolus is caused by *Xanthomonas gummisudans* in wet weather, or when plants are growing in poorly drained soil. It begins as dark green, translucent, watersoaked horizontal spots. As spots enlarge, they become rectangular brown patches. Bacterial slime can form a sticky coating on leaves. Another bacterial disease, caused by *Burkholderia gladioli* pv. *gladioli*, is known as scab. Scab symptoms are brown leaf spots and corms that show sunken brown crater-like lesions with raised edges, accompanied by a shiny, brittle bacterial exudate. Soft rot at the stem base may cause plants to fall over. To manage bacterial diseases, handle plants only when dry, and avoid excessive nitrogen application or irrigation methods that leave foliage wet for long periods. Inspect corms before planting, and destroy plants if scab is detected. Shoot proliferation caused by the bacterium *Rhodococcus fascians* also occurs on gladiolus.

Botrytis gladiolorum and *B. cinerea* affect flowers, leaves, stems and corms. These fungi are carried to new areas as sclerotia on corms. The brown leaf spots are of assorted sizes and may have reddish borders. Flower petal spots are initially translucent and then turn brown, after which the dead tissue may be covered by the gray mold formed as the fungus sporulates. Corms may show brown rotted areas and bear black sclerotia on the surface. Check the corms for these sclerotia before planting to avoid installing diseased plants. Choose planting sites with good drainage and air movement.

Soft brown spots on gladiolus corm due to *Burkholderia gladioli.*

Shoot proliferation on gladiolus caused by *Rhodococcus fascians.*

The rust disease of gladiolus caused by *Uromyces transversalis* is not established in the United States, but it was observed in Florida and California in 2006-8. Plants develop rust pustules on the undersurface of the leaves, aligned perpendicular to the main vein of the leaf (hence the name "transversalis"). If you suspect that you have this disease in your gladiolus planting, report the disease to a horticultural inspector with your state department of agriculture. The other hosts of this rust are *Crocosmia*, *Freesia*, *Tritonia* and *Watsonia* spp. A smut disease cause by *Urocystis gladiolicola* also occurs occasionally on gladiolus.

Small, round, reddish lesions on leaves are the indication of Stemphylium leaf spot, caused by the fungus *Stemphylium botryosum*. Varieties vary in their susceptibility. In contrast, Curvularia leaf spot symptoms are diamond-shaped tan spots that have brown borders, surrounded by a yellow halo. One of the fungi causing these leaf spots, *Curvularia lunata*, can also rot the corms. Both Stemphylium and Curvularia leaf spots are more of a problem in unusually wet years. Gather up plant debris in the fall to reduce carryover of the pathogens.

Neck rot on gladiolus may be due to Stromatinia dry rot, another corm disease. This fungus problem appears as small red-brown spots on the corm surface that become corky, sunken, round areas with raised edges. Some corms become hard, dry, and mummified. Tiny black sclerotia form on leaf bases and within the dead areas on corms. Be careful to plant only sound corms. Unfortunately, the sclerotia of *Stromatinia gladioli* can live for many years in infested soil, so glads should not be replanted in an infested area for at least 7 years. *Rhizoctonia solani* may also cause a neck rot on gladiolus.

Dark brown discolored spots caused by *Botrytis gladiolorum* on gladiolus corms.

Fusarium wilt is caused by a host-specific form of *Fusarium*, *Fusarium oxysporum* f. sp. *gladioli*. The disease is often recognized by yellowing foliage and overall plant stunting, as well as failure to flower. Flower stalks may be curved. Corms show both internal streaks of discoloration

Extensive discoloration on gladiolus corms due to *Fusarium oxysporum*.

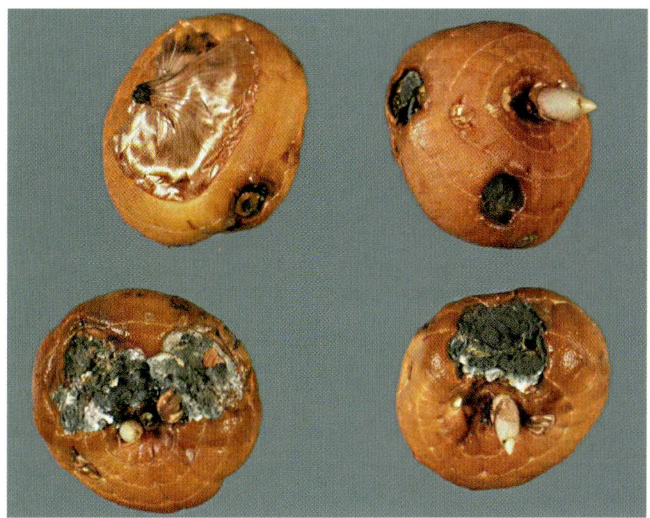

Bluish sporulation of a *Penicillium* sp. on gladiolus corms.

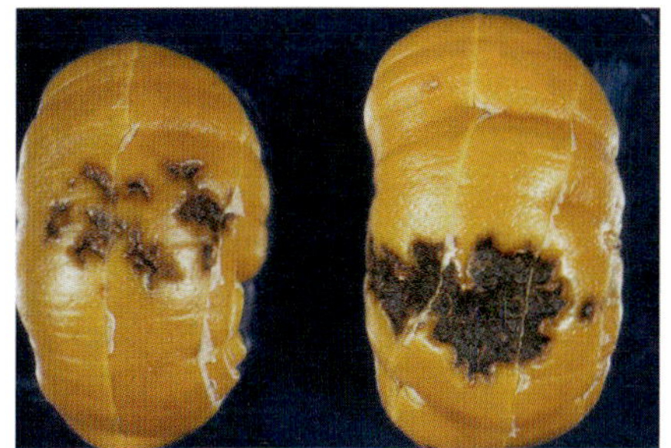

Brown pitted lesions caused by *Septoria gladioli* on gladiolus corms.

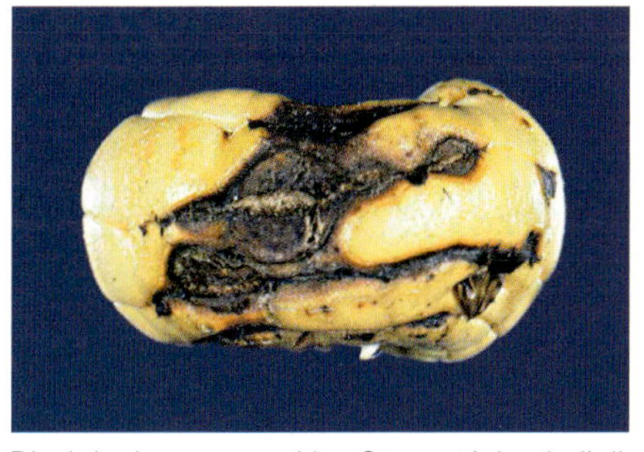

Black lesions caused by *Stromatinia gladioli* on a gladiolus corm.

Gladiolus corm discoloration due to *Ditylenchus destructor*.

Flower break in a pink-petaled gladiolus due to *Cucumber mosaic virus* (CMV).

and brown spots on the surface, with rot beginning at the basal plate. Irregular concentric rings, indicating periods of expansion, are visible within the corm lesions. Affected corms may develop dry rot during storage. Varieties vary in susceptibility; try to grow resistant varieties. Cormels can be given a hot water treatment to eliminate *Fusarium*. Keeping the soil pH at 6.6 to 7.0 and utilizing nitrate forms of nitrogen are good practices for counteracting Fusarium wilt in plants. Setting up a new planting bed with new, clean material is ideal, if there is space in the garden to make a fresh start.

Virus symptoms are also sometimes seen in gladiolus: flowers may be stunted, distorted or show color break; foliage may show mosaic, white flecks, or red blotches. Viruses that infect gladiolus include *Cucumber mosaic virus* (CMV), bean yellow mosaic (BYMV) (also known as *Gladiolus mosaic virus*), tomato ringspot (ToMV) and tobacco ringspot (TRSV). The aphid-vectored BYMV causes a faint mosaic, mottling and sometimes flower color break in gladiolus. The virus is spread both by aphids and mechanically during harvest. ToMV and TRSV cause light-colored ring patterns and blotches on leaves; these viruses are spread by nematodes as well as mechanically during harvest. CMV causes white break of gladiolus, in which the affected petals show white blotches or color break and flowers may not open normally; it is spread primarily by aphids. *Gladiolus latent virus* (GLV) is also known as iris mild mosaic and *Iris mild yellow mosaic virus*: it is usually asymptomatic in gladiolus.

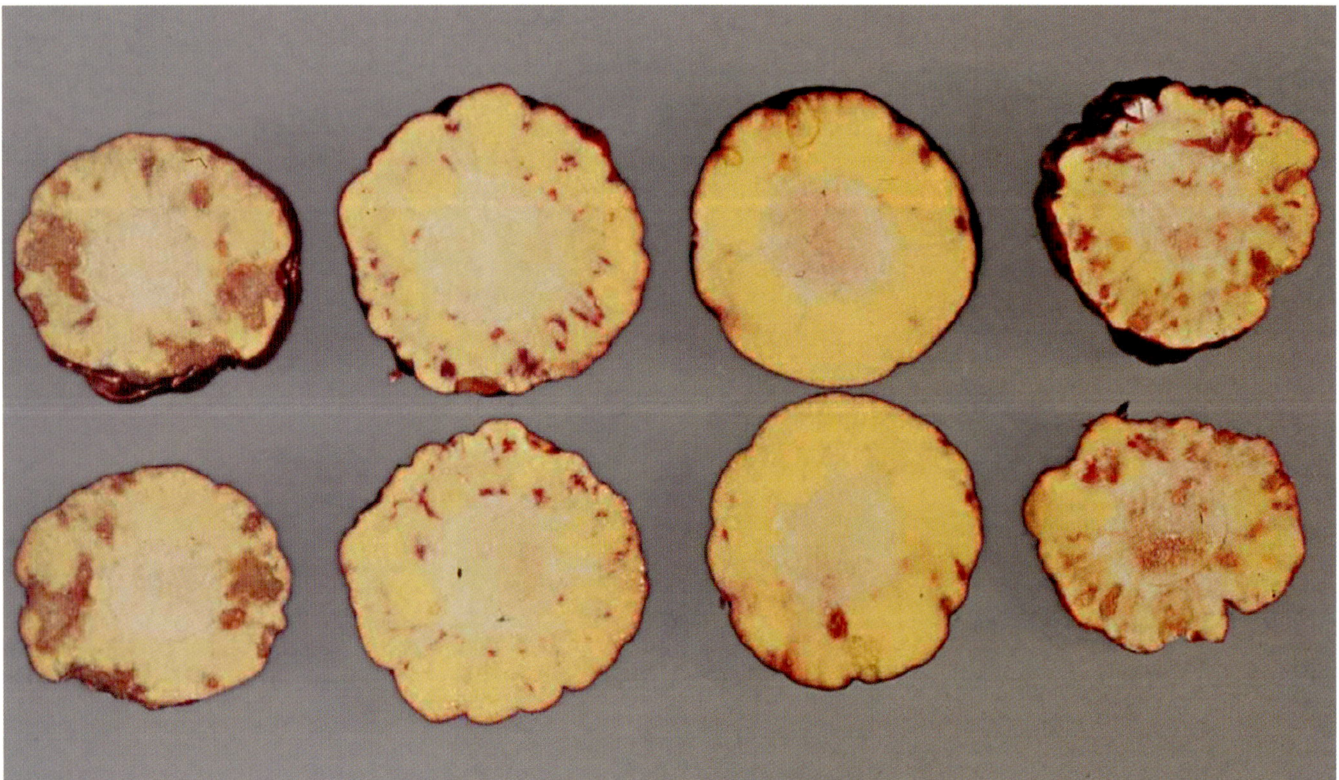

Purplish discoloration associated with *Cucumber mosaic virus* (CMV) infection in gladiolus corms.

Gladiolus is also susceptible to aster yellows, which leads to a condition called grassy top. Plants with aster yellows may be stunted and produce green flowers on twisted flower spikes. The thin weak shoots produced the year after infection lead to the name grassy top. Controlling leafhoppers that vector the disease and weed hosts in the vicinity of the cultivated plants will reduce the impact of aster yellows. Remove infected plants, including the corms, as soon as they are noticed.

For garden use, it is a good practice to start with healthy new corms each year. If storing your own corms, you should inspect them in the spring and discard those with any signs of discoloration or rot before planting. Commercially, corms must be cured after digging, cleaned, stored, given a preplant dip in fungicide plus wetting agent, and dried before planting.

Symptoms of *Tobacco rattle virus* (TRV) on gladiolus leaves.

Petal lesions caused by *Botrytis* sp. on gladiolus.

Stunting, wilting and death of gladiolus plants due to Fusarium wilt.

Sporulation of the rust *Uromyces transversalis* on gladiolus.

Stemphylium leaf spot on gladiolus.

Grasses and Sedges, Ornamental

Poaceae and Cyperaceae

A design trend has immensely increased the use of ornamental grasses in perennial gardens during the past few decades. One reason for their great popularity is that these plants generally have few disease problems. In some cases disease symptoms appear on the foliage, but are not unsightly at a distance, and thus infected plants can be tolerated in the landscape. Some of the key genera of grasses used in perennial gardens are *Calamagrostis*, *Hakonechloa*, *Miscanthus*, *Panicum*, *Pennisetum* and *Phalaris*. Sedges, *Carex* species, are also used frequently in modern landscapes.

Most grasses and sedges do well in full sun locations, in moist soil of moderate fertility. Winter injury may be a problem in excessively wet sites. Excessive moisture on the foliage of any of these ornamental grasses and sedges will increase the extent of disease symptoms. Many of the same genera that are found on turfgrasses will affect the larger ornamental specimen grasses. To reduce the risk of foliage diseases on these plants, water clumps at the base rather than by overhead irrigation, and water in late morning to midday if overhead irrigation is the only option. In nursery production where watering is more difficult to regulate, fungicides may be needed to supplement cultural control.

FOUNTAIN GRASS

Fungal diseases reported on fountain grass (*Pennisetum* spp.) in the United States include anthracnose (*Colletotrichum* species), leaf spot (caused by *Bipolaris*, *Cercospora* and *Phyllosticta* species), seed smut (*Ustilago* species), and black choke disease (*Ephelis* species), along with *Maize chlorotic dwarf virus* (MCDV). Crown rots are due primarily to *Rhizoctonia* and *Sclerotium* species. Again, these problems are most likely under excessively wet cultural conditions, or when plants are overly shaded. Good drainage and sun exposure will improve their chances of overwintering in more northern gardens in the United States.

HAKONE GRASS

Healthy hakone grass.

Hakone grass (*Hakonechloa* spp.) is susceptible to an anthracnose caused by a *Colletotrichum* species as well as Pyricularia leaf spot and other foliar fungal diseases. Overhead watering should be minimized with variegated forms, as these are more prone to leaf diseases.

Pyricularia leaf spot on hakone grass.

MAIDEN GRASS

Healthy maiden grass.

Maiden grass (*Miscanthus sinensis)* may be affected by a number of leaf spotting fungi, including *Bipolaris*, *Curvularia*, *Colletotrichum*, *Stagonospora*, and *Stemphylium* species. The most dramatic fungal leaf blight on this host is a disease due to a *Stagonospora* species called miscanthus blight. Purplish or rust-colored spots and streaks occur on both leaves and sheaths. Tips and margins of leaves are reddened, purpled or killed—in some cases entire plants are stunted or killed. The rusty discoloration is commonly seen in the white areas of variegated leaves. Although the rusty color causes some to mistakenly refer to this as a rust disease, the cause is not a rust fungus. Spore structures of both *Stagonospora* sp. and *Leptosphaeria* sp., the sexual stage of the fungus, are visible on the dead areas. Other leaf spotting symptoms on miscanthus may be due to anthracnose (*Colletotrichum* species). Stunted, bunched foliage and abnormal flowers on miscanthus may be the result of mealybug infestation: look for these culprits hidden between the leaf sheath and the stem. Miscanthus plants may also occasionally show a true rust disease, but the symptoms are more subtle than those of miscanthus blight. To identify a rust, look for spore masses produced in pustules on the leaf undersurface.

A true rust disease on miscanthus.

Miscanthus blight caused by *Stagonospora* causes tip dieback, purplish leaf streaks or rust-colored discoloration.

REED GRASS

Healthy reed grass.

Reed grass (*Calamagrostis* spp.) is prone to rust (*Puccinia* spp.) and anthracnose (*Colletotrichum graminicola*). The disease ergot, caused by *Claviceps purpurea*, converts tissue that would normally have developed into a seed into a hard black fungal sclerotium, which is larger, longer and darker than a healthy seed. The black sclerotia are conspicuous within the seedhead of the affected grass plant. The plant is not significantly harmed overall, although its reproduction is reduced by the action of the fungal parasite.

SWITCHGRASS AND MILLET

Healthy switchgrass.

Garden varieties of switchgrass (*Panicum virgatum*) and millet (*P. miliaceum)* are drought tolerant. Leaf spot and blight diseases, often caused by species of the fungi *Alternaria, Bipolaris, Colletotrichum, Curvularia, Phyllachora, Leptosphaeria, Septoria,* and *Phyllosticta,* may appear on switchgrass during wet growing seasons but rarely affect plant vigor. Other widely occurring diseases on panicums include rust (caused by *Puccinia* species and *Uromyces graminicola*), smut (*Tilletia maclaganii*), head scab (*Giberella*) and root and stem rots caused by *Rhizoctonia solani* and *Fusarium* spp. The rust diseases are especially damaging to some of the blue panicums. The cultivars 'Rotstrahlbusch', 'Rotbraun', and 'Shenandoah', for example, have been less rust-prone than 'Heavy Metal', 'Warrior', and 'Prairie Sky'.

REED CANARYGRASS

Healthy reed canarygrass.

Reed canarygrass *(Phalaris* spp.) can tolerate wet sites, but will show healthier foliage if not exposed to frequent overhead irrigation. It is reportedly attacked by a range of leaf spot fungi, including *Ascochyta*, *Helminthosporium* and *Stagonospora*, as well as rusts (*Puccinia*), ergot (*Claviceps purpurea*), and root and stem rot pathogens (*Pythium* and *Rhizoctonia solani*).

SEDGES

For some sedges, bronze foliage is normal.

Sedges (*Carex* spp.) may develop Pythium root rot if overwatered. Leaf spot diseases are known for many species of carex. Fungal pathogens of carex foliage include species of *Cladosporium*, *Mycosphaerella*, *Phyllosticta* and *Septoria*. A Uromyces rust has been reported from Canada on *Carex flacca*, and many different Puccinia rusts have been reported from a wide range of sedge species.

Black sclerotia of ergot within the seed heads of feather reed grass.

Rust symptoms and signs on switchgrass (*Panicum* sp).

Rust pustules on the underside of blue sedge (*Carex flacca*) leaves.

Gypsophila

Caryophyllaceae

Gypsophila, or baby's breath, includes perennials producing sprays of little flowers that are known for their contribution to bouquets, both fresh and dried. Most garden types are varieties of *G. paniculata*. These require sturdy staking because they become quite top-heavy with a profusion of white or pink flowers. In southern regions, cutting the plant back after flowering will allow it to rebloom. Plants have deep taproots, so they will not transplant easily. *Gypsophila* spp. do best in moist, well drained soils at neutral or slightly alkaline pH, in full sun. Mulch once the ground has frozen to protect from winter injury, and avoid winter moisture in the garden.

Flowers of gypsophila may be blighted by the fungi *Botrytis* or *Alternaria*. With Botrytis blight, the disease can progress until flower stems turn ashy-gray as they are killed. Keep plants well spaced and avoid overhead watering to reduce Botrytis symptoms. Leaf spots of gypsophila may be due to *Alternaria*, *Botrytis* or *Phyllosticta* species. Several different species of *Phytophthora* and *Pythium* are known to affect gypsophila; these water molds cause root and crown rots that lead to wilting or stunting of the plants, and the affected plants are easily pulled out of the ground. The fungus *Rhizoctonia solani* also may attack the stem near the soil line, and cause wilting of the whole plant, or sometimes only one side of the plant where the canker is located. In other cases, stem rot may be due to *Fusarium* species. Avoid deep planting because this practice favors crown rots due to a wide range of pathogens. A bacterial gall disease caused by the bacterium *Pantoea agglomerans* pv. *gypsophilae* (previously known as *Erwinia herbicola*) may occur on gypsophila: if plants appear stunted, check at the soil line or on the root systems for small, soft brown galls, up to 2 inches diameter. Discard plants with this symptom.

Wilting of baby's breath due to Rhizoctonia stem rot.

Galling at the root crown of baby's breath caused by infection with the bacterium *Pantoea agglomerans* pv. *gypsophilae*.

Fusarium flower blight on baby's breath.

Helenium

Compositae

Helenium, or sneezeweed, includes ornamentals suitable for sunny gardens, in moist, fertile soil. This plant is not attractive if grown in dry soil. Many of the garden hybrids are crosses between *H. autumnale*, native to the east and north central United States, and *H. bigelovii*, native to the West Coast. Bloom is improved by deadheading; plants are most vigorous when divided every 2 or 3 years in either fall or spring. Cutting them back mid-season is advisable in the South.

Helenium is prone to powdery mildew, a fungus disease that will form a white coating of fungal mycelium and spores across the leaf surface. White smut diseases caused by *Entyloma* species are also possible: these fungi form yellow spots that gradually turn brown. Helenium can get several kinds of fungal leaf spot, including ones caused by *Ascochyta*, *Cercospora*, *Cladosporium* and *Septoria* species, but these are not very common. Rusts have also occasionally been reported on sneezeweed. *Helenium virus S* (HVS) is a carlavirus that causes no symptoms in helenium. In contrast, *Helenium virus Y* (HVY), a potyvirus found in Germany that may occur together with HVS, is associated with flower break and leaf yellowing.

Powdery mildew on sneezeweed.

Helianthemum

Cistaceae

Helianthemum, or sun rose, is a subshrub. It is grown in sunny locations, in neutral to alkaline soil that is very well-drained. The species *H. nummularium* is grown in the United States. This species should be cut back after it flowers. It needs partial shade in southern gardens and will do better in cooler climates. Hybrids are also available.

Helianthemum is subject to few foliar diseases. The plants are, however, able to contract a web blight caused by *Rhizoctonia solani* that will kill foliage under warm, humid growing conditions. Wet soil conditions may make the plants prone to Pythium root rot.

Helianthus

Asteraceae

Helianthus, or sunflower, includes species that grow robustly and can be invasive. The garden plants are relatives of the seed crop, *Helianthus annuus*, and *H. tuberosum*, the Jerusalem artichoke. For best performance, sunflowers are grown in full sun, in moist, neutral to alkaline soil of average fertility that drains well. Some are quite drought tolerant, but *H. angustifolius* needs plenty of water. Fertilization is generally beneficial. Plants should be divided every 2-4 years.

The most widespread disease on helianthus is undoubtedly powdery mildew, which affects the plants by midsummer. Small white blistery lesions on leaves of sunflowers are symptoms of a white rust disease, caused by the oomycete *Albugo tragopogonis*. Downy mildew caused by *Plasmopara halstedii* can cause chlorosis at the base of the leaves and along the midvein. White downy mildew sporulation appears on the undersurface. Alternaria leaf spot, caused by a fungus, and Xanthomonas leaf spot, caused by a bacterium, are the most common leaf diseases on sunflower. Both of these have an angular shape. Additional fungal diseases include Ascochyta, Botrytis, Cercospora, or Phyllosticta leaf spots and anthracnose caused by a species of *Colletotrichum*. A number of different species of rust (*Puccinia*, *Uromyces* or *Coleosporium* species) are very common on sunflowers.

Fungal cankers on stems may be due to infection by a *Diaporthe* species. Plants that wilt suddenly might be cankered by *Diaporthe* or might be under attack by a *Fusarium* sp. at the stem base, or by a vascular wilt caused by *Fusarium* or *Verticillium* species. Bacteria (*Pectobacterium)* are also sometimes responsible for stem rot of helianthus. Root rot or stem rot problems due to *Pythium*, *Phytophthora*, or *Rhizoctonia* are also reasons for plant stunting or sudden wilt.

Sclerotinia sclerotiorum and *Sclerotium rolfsii* can both cause death of the root crown; the shape and size of the sclerotia formed by these two fungi will allow you to tell them apart. *Sclerotinia sclerotiorum* has large black sclerotia resembling mouse droppings whereas *S. rolfsii* has smaller tan to purplish brown, round sclerotia that look more like mustard seeds. Galls on the root system or at the stem base may be due to crown gall disease, caused by a bacterium, *Agrobacterium tumefaciens*. This bacterium

Brown canker at the soil line caused by *Rhizoctonia solani* on a sunflower stem.

White mycelia and black sclerotia of *Sclerotinia minor* on sunflower stems.

Patches of white rust (*Albugo candida*) sporulation on sunflower.

Powdery mildew on sunflower.

Rust on sunflower (upper leaf surface).

Rust on sunflower (lower leaf surface).

may contaminate soil and cause new infections via wounds to the root or stem of a wide range of plants that serve as hosts. Affected plants may be stunted or relatively prone to winter injury.

There are only a few viruses reported to affect *Helianthus* spp. naturally in the United States: *Cucumber mosaic virus* (CMV), *Tobacco ringspot virus* (TRSV), and an ilarvirus that causes sunflower necrosis disease. *Helianthus mosaic virus*, also known as *Sunflower mosaic virus* (SuMV), has been noted in wild *H. annuus* in Texas; it causes a systemic mild mosaic and mottle, after which leaves and stems may turn brown and dry. *Bidens mosaic virus* (BiMV) has been found causing mosaic and necrosis on *H. annuus* in Brazil.

Frosty patches of downy mildew sporulation on the undersurface of a sunflower leaf.

Heliopsis

Asteraceae

Heliopsis is a genus of North American natives also known as false sunflower. These are smaller relatives of the true sunflowers, *Helianthus*. Heliopsis are suited to full sun locations, and should be fertilized only lightly. The species *H. helianthoides* tolerates dry conditions and is not particular about soil type. The varieties do best in well-drained fertile soils high in organic matter; moisture should be supplied during droughty periods. Plants should be deadheaded to prolong blooming, and divided every 2-3 years to keep them vigorous.

Powdery mildew fungi form white colonies on the leaves that coalesce over time. Coleosporium and Puccinia rusts occur fairly often. Fungal leaf spots due to *Ascochyta*, *Septoria*, *Cercospora*, and *Phyllosticta* are more occasional.

Water-soaked Pseudomonas leaf spots on heliopsis.

Powdery mildew on heliopsis foliage.

Mosaic-like pattern of injury on heliopsis leaves infested with foliar nematode.

Helleborus

Ranunculaceae

Helleborus, or hellebore, has petal-like sepals rather than showy petals. These plants prefer partly shaded sites, rich in organic matter. Hellebores grow best in neutral or alkaline soil in most cases, although some prefer slightly acid conditions. They are naturally suited for woodland gardens. The plants are drought-tolerant once they are well rooted. They should be sheltered from winter winds, and soil moisture extremes should be avoided. Hellebores do not appreciate disturbance, and are long-lived; rather than propagating them by division, seedlings are generally collected from the base of plants. Hellebores are poisonous, and the sap may cause skin irritation.

Hellebores are subject to winter leaf injury due to the effect of cold and desiccation on their exposed foliage. Contagious diseases may also contribute to the brown dead areas on the leaves, particularly leaf spots due to the fungus *Coniothyrium hellebori*, which causes black leaf spot on *H. niger*. An unidentified downy mildew has also been observed on this plant. Root rots may be caused by *Pythium* species, and stem and crown rot problems may be due to the fungi *Rhizoctonia*, *Sclerotinia* or *Sclerotium delphinii*. Excess moisture on foliage or poor drainage in nursery containers or the landscape will encourage these diseases. Fungicide treatments may be helpful in production areas. In the garden, remove plants with root or stem diseases and replace with different species.

Anthracnose leaf lesions caused by *Colletotrichum* sp. on hellebore.

Tomato spotted wilt virus (TSWV) may cause curious looking black line patterns in the foliage. This virus is introduced by feeding of the western flower thrips, an insect so tiny that it often escapes notice until virus symptoms appear. Thrips are a particular problem during greenhouse propagation. Hellebores are also susceptible to *Cucumber mosaic virus* (CMV), which may cause mottling and rugose foliage, as well as to *Chrysanthemum virus B* and *Helleborus mosaic virus*. A hellebore disease referred to as "black death" has been of recent concern in North America as well as Europe. Symptoms reported on *Helleborus* x *hybridus* are seen primarily on new foliage and flowers of mature plants: blackening along the leaf veins, dark streaks or ring spots on petioles and bracts, and stunted, blackened new growth. Hellebores with "black death" are severely stunted, and may die rapidly or gradually, over the course of a few growing seasons. "Black death" is caused by a carlavirus with the proposed name Helleborus net necrosis virus (HeNNV). Transmission of black death by *Macrosiphum hellebori* (Helleborus aphid) has been demonstrated. Hellebores with symptoms of this disease should be removed from the nursery or garden immediately. Aphid management is critical in nurseries to prevent spread.

Phyllosticta leaf spot on hellebore.

Hellebore with Pythium root rot.

Cloudy necrotic areas in hellebore leaves with downy mildew.

Leaf-edge necrosis caused by *Xanthomonas* sp. on hellebore.

Watersoaked appearance of hellebore leaves infected by downy mildew.

Black line pattern caused by *Tomato spotted wilt virus* (TSWV) on hellebore.

Botrytis stem rot on hellebore, showing a canker covered with the characteristic gray mold sporulation.

Blackening along veins is one of the symptoms of "black death", due to infection of hellebore by a carlavirus.

Vein-bounded patches of discoloration on hellebore leaves caused by foliar nematode.

Coniothyrium leaf spots on hellebore.

Hemerocallis

Liliaceae

Hemerocallis, or daylily, is a popular garden plant for both private and public garden areas across the United States. Species daylilies are not seen as often as hybrids, of which there are thousands. These plants like sun or partial shade and well-drained soils. They do best in rich soils with plenty of organic matter but can be grown in a wide variety of soils. Blooming is strongest when daylilies are watered during summer dry periods and when shade is not excessive. Fertilize during spring; excess nitrogen will lead to more leaves and a weaker flower display. Divide plants every 3-5 years. Thrips, mites and aphids can be a problem. Heavy infestations of daylily thrips may cause distortion of the flower buds.

Hemerocallis are considered relatively pest free. The most common fungal diseases are rust (*Puccinia hemerocallidis*), leaf streak (*Aureobasidium microstictum*), crown and root rot (*Sclerotium rolfsii*), and root rots (*Armillaria, Rhizoctonia)*. Species of *Pythium* also cause root rots, and a *Pectobacterium* species causes a bacterial soft rot.

Symptoms of daylily rust appear in early summer, and are initially seen as tiny yellowish flecks on the leaves. Rust is conclusively identified by seeing the raised pustules, most common on the lower leaf surface and sometimes on the upper leaf surface or on scapes of affected plants: these pustules bear bright orange colored spores. Very susceptible varieties should be avoided in gardens and nurseries so that fewer spores will be produced. If rust is introduced to a garden or nursery, remove, then burn or bury leaves from infected plants, as well as plants in the same area, after harvesting the top growth as close to the soil as possible. Fungicides may be sprayed on the new growth as it emerges to prevent new infections. The goal is to eradicate the rust from the area. If rust becomes established in a garden or nursery, only the most rust-resistant varieties will be able to be grown without a strict fungicide program.

Daylily leaf streak is present most places where daylilies are grown. It is due to the infection of *Aureobasidium microstictum*, a fungus. Symptoms begin to show in early spring and first appear as small water soaked spots on the leaves. These spots gradually become brown streaks with yellow borders that usually develop from the leaf tip

Daylily leaf streak caused by *Aureobasidium microstictum.*

downward; infected leaves may die prematurely. The fungus requires leaf damage (from frost, thrips feeding, etc.) in order to infect daylily. To manage daylily leaf streak, purchase disease-free stock plants and propagate only from healthy specimens. Also, avoid highly susceptible varieties, remove infected leaves, and limit overhead irrigation.

Spring sickness is a disorder that causes fans to grow slowly in the spring. Other symptoms include sideways growth with brown ragged edges and holes and decay at the base. Fans affected by spring sickness remain stunted and do not bloom, although many do resume growing and produce scapes. The cause of spring sickness is unknown and has been attributed to freeze damage, slugs, genetics, fungal infection, a nutritional disorder, and feeding injury from thrips, tarnished plant bugs, bulb mites, and lesser bulb flies.

Fungi and oomycetes that cause crown and root rots include *Armillaria*, *Rhizoctonia*, *Pythium*, *Phytophthora*, and *Sclerotium* species. The prevention of root rot diseases often depends on careful water management to avoid root stress. Symptoms of root stress often resemble nutrient deficiencies, as well as stunting and overall poor plant growth. *Pythium* and *Phytophthora* are favored by the same conditions that reduce root growth, such as high soil moisture, high soluble salts, and poor aeration. *Rhizoctonia solani* is favored by stress conditions, including excessive drying and rewetting of the soil medium. In general, the longer the soil stays wet, the greater the likelihood of developing root problems. A well-aerated soil with good drainage, in contrast, will reduce root diseases. Armillaria root rot is unlikely to occur in container production nurseries, but may be encountered when freshly

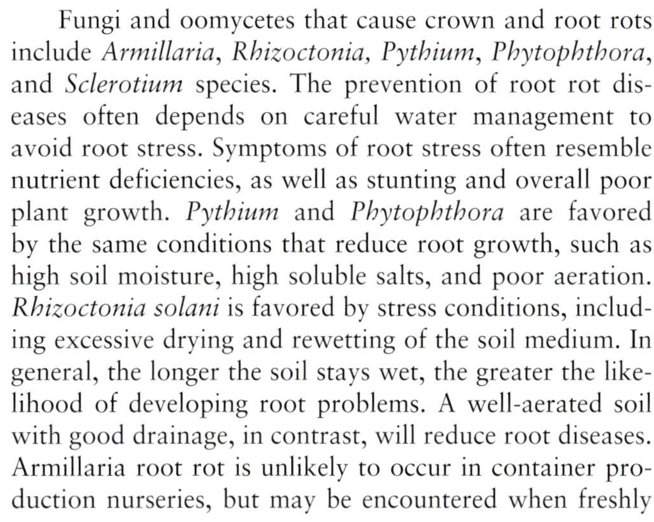

Closeup of daylily rust, showing orange spores spilling out from pustules.

Daylily rust caused by *Puccinia hemerocallidis*.

Symptoms and signs of daylily rust on *Patrinia*, its alternate host plant.

Spring sickness of daylily is a disease of unknown cause; symptoms include stunting, spotting and tattering of early growth.

cleared land is used for growing daylilies. Plants will show low vigor, and inspection of the rhizomes will show rotted areas and black shoelace-shaped fungal structures called rhizomorphs.

A *Pectobacterium* species can cause soft mushy rot at the base of flowers and in rhizomes. To prevent bacterial soft rot, avoid poor soil drainage by adding organic matter to heavy clay soils and avoid poor air circulation conditions in nursery and garden plantings.

Root-knot nematodes (*Meloidogyne*) cause stunting, yellowing, wilting, and premature death. Roots are swollen with small bumps. In garden areas contaminated with root knot nematode, it will be best to grow only plants that are not hosts for *Meloidogyne*.

Collapse of lower foliage of daylily due to a Phytophthora crown rot.

Rhizomorphs of *Armillaria* sp. intertwined with white roots of daylily.

Wilting and necrosis of daylily foliage caused by Armillaria root rot.

Chlorotic streaking and spotting due to *Tobacco ringspot virus* (TRSV) in daylily.

Heuchera

Saxifragaceae

Heuchera, or alumroot, is a long-lived perennial genus native to North America and Mexico. These plants are grown for both foliage and floral display. There are many species and varieties, including the familiar coral bells, *H. sanguinea*. Heucheras do best in moist, well-drained soil, rich in organic matter, at a pH near 7.0. Partial shade or sun is preferable; heucheras do not do well in dry shade. Replanting is needed after a number of years to reset the crowns just above the surface of the soil. Ability to tolerate heat and humidity varies among species and varieties, so selection of the most appropriate heucheras is especially important for southern gardens. Foliage may turn dull from the heat of midsummer. In areas that experience frost, mulch for the winter to minimize heaving. Divide heucheras every 4-5 years. *Heucherella* or Foamy Bells are intergeneric hybrids that were formed by crossing *Heuchera* with *Tiarella*. Their culture is very much like that of *Heuchera*. Heucherellas are short-lived in the South, but do well in other regions of the United States. These plants grow best in moist, partly shady gardens in light, well-drained, neutral to acidic soil; they can tolerate full shade. Heucherellas are susceptible to many of the diseases hosted by heucheras and tiarellas.

The most serious contagious problem on heucheras is the rust disease caused by *Puccinia heucherae*. The undersurface of leaves may be covered with brown rust pustules. Severe symptoms have been noted under nursery production conditions, when overhead watering supplies abundant moisture for the spread of the rust fungus.

Heuchera sanguinea is susceptible to three different kinds of pathogens that may cause bothersome leaf spots: fungi, bacteria and nematodes. The fungal leaf spots are typically due to *Alternaria*, *Botrytis*, *Cercospora*, *Septoria*, *Ramularia*, *Pestalotia* or *Colletotrichum* species. The bacterial leaf spot is caused by a *Xanthomonas* species. The lesions it causes are distinguished from the fungal leaf spots by the absence of any fungal sporulation on spots kept in a moist chamber for a few days. In garden situations, pick off heuchera leaves with leaf spots; in

Rhizoctonia crown rot of heuchera.

Pestalotia leaf spots on heuchera.

Ramularia leaf spot on heuchera

commerical production, fungicides may be used. Be sure to use bactericidal materials if you are not sure whether spotting is fungal or bacterial in nature. Copper materials are a good choice in these situations, as they curtail the growth of both bacteria and fungi. If leaf spotting is due to foliar nematode (an *Aphelenchoides* species), the spots would also show no sporulation, and neither fungicides nor bactericides would be of any benefit. A foliar nematode diagnosis may be made only by putting pieces of leaf material into a dish of water and examining them several hours later with the aid of a microscope for the presence of the tiny worms, which will swim in the water with a serpentine motion. Reduce foliar wetness intervals if foliar nematode is encountered. Since many other species of perennials are also susceptible to foliar nematode, discarding plants with this problem will protect the overall health of plants in a garden or nursery.

Root rot in heuchera is likely to be due to a *Pythium* or *Phytophthora* species, and to be brought on by overly wet soil conditions.

Turnip mosaic virus (TuMV) causes a yellow mosaic in heuchera leaves. *Tobacco rattle virus* (TRV) has also been detected on coral bells. This virus may be vectored by certain species of root-feeding nematodes, so if plants with this virus are removed from the garden they should be replaced with a different species not known to be susceptible to TRV.

Xanthomonas leaf spot on heuchera.

Botrytis leaf spot on heuchera.

Brown spotting due to foliar nematode feeding in heuchera leaves.

Rust pustules on the undersurface of a heuchera leaf.

Tobacco rattle virus (TRV) on heuchera.

Bacterial leaf spot on heuchera.

Hibiscus

Malvaceae

Hibiscus, or Confederate rose, includes shrubs and trees as well as herbaceous plants. The hibiscus grown as a perennial is *H. moscheutos*, the common rose mallow (this plant has also been referred to as *H. palustris*). Native to marshy locations, a good supply of moisture is needed for rose mallow to thrive (one of its common names is swamp mallow). Some varieties are crosses with *H. coccineus* and *H. militaris*. All of these are native to the United States.

The large leaves of the rose mallow are subject to infection by a rust, *Puccinia schedonnardi*, and also to leaf spots caused by *Phyllosticta hibiscina* and *Cercospora* and *Septoria*. Many additional leaf-spotting fungi have been reported from its tropical ornamental relative, *Hibiscus rosa-sinensis*. Rose mallow is susceptible to a bacterial leaf spot caused by a *Xanthomonas* species. The fungi *Fusarium* and *Sclerotium rolfsii* may cause stem and crown rot symptoms, while *Phymatotrichopsis omnivora* may cause root rot in the southern United States. *Hibiscus* spp. are also susceptible to crown gall caused by the bacterium *Agrobacterium tumefaciens*. The disease causes swellings that form at the stem base or on the roots. Crown gall has a very wide host range, so it is important to examine plants for galls at the soil line before purchase.

Alternaria leaf spot on hibiscus.

Xanthomonas leaf spot on hibiscus.

Hibiscus basal stem canker caused by a *Fusarium* species.

Hosta

Liliaceae

Hosta, or plantain lily, is a group of long-lived Asian natives especially well suited to shady gardens. They are grown largely for their foliage. Hostas grow well in humus-rich, moist, well-drained soils in partial shade, and may tolerate light levels varying from full sun to considerable shade. Hundreds of forms of hosta are available, many of them varieties or hybrids of multiple species. Drought injury results in smaller leaves with papery dead areas along the margins. Predation by voles, deer, black vine weevils and slugs can be very injurious to hostas. High temperature and high nitrogen may cause a loss of variegation.

Anthracnose, caused by several fungi in the genus *Colletotrichum*, is the most widespread foliar disease of hosta. Disease is favored by moisture and warm temperatures. Anthracnose causes large, irregular spots with darker borders. Centers of spots often fall out, and leaves become tattered. Cultural practices to minimize this disease include regular irrigation, keeping plants cool, and removing infected leaves. Fungicides can protect new growth, but should be applied before symptoms appear. Less common are fungal leaf spots caused by *Cercospora* and *Botrytis* species and a foliar blight due to a *Phytophthora.*

Petiole rot, caused by the fungi *Sclerotium rolfsii* and *S. delphinii*, can rapidly damage or kill hosta plants and is difficult to eradicate. Small spheres produced by the fungus, called sclerotia, allow the fungus to survive cold or dry periods. When warm, rainy weather occurs and hostas are nearby, sclerotia become active, produce stringy white threads (mycelium) that infect the plants, and soften and kill the petioles at the base. Once petiole rot has entered a garden or nursery planting it is easily spread by sclerotia clinging to soil on shoes, tools, and plant material. Lower leaves begin to turn yellow and brown, then wilt. Affected petioles are mushy and brown at the base. Mycelia and sclerotia of the fungus show up on the rotted tissue and surrounding soil. It's helpful to inspect plants carefully before planting, and discard and destroy affected plants. If the disease has already appeared, control becomes more difficult, because sclerotia can survive for years. Garden areas with infected hostas and soil should be quarantined, and care should be

Both *Alternaria* and *Fusarium* spp. have contributed to the leaf spotting on this hosta.

Interveinal browning is a symptom of foliar nematode on hostas.

used to avoid spreading the soil or plant material outside of this zone. Fungicides can be applied preventively to soil or growing media.

Fusarium root and crown rot is caused by several related species of fungi that live in the soil. *Fusarium hostae* invades hosta plants through wounds, especially under dry soil conditions when temperatures are between 60 and 80°F. Leaves turn yellow, then tan, and wither. Below ground, there are fewer and shorter roots. Roots close to the crown show brown to black discoloration. Diseased plants often are stunted and slow to emerge in spring. Diseased plants should be removed and destroyed. A good preventive measure is to minimize wounding to the roots and crowns when plants are transplanted. Divided plants can be dipped in a fungicide suspension before planting. Plants growing in pots or in the landscape should be watered regularly.

Bacterial soft rot of hosta can be caused by several *Pectobacterium* species. Soft rot often follows damage by ice or freezing temperatures. Dividing plants with knives or manually can also spread the bacterium. Infected tissue becomes watery and soft, with a foul smell. Leaves turn yellow and wilt. A soft rot at the base of petioles may cause plants to collapse and die. Strict sanitary practices are important in managing bacterial diseases; when separating plants, discard any that have symptoms, and clean all knives or other nursery tools, hands, and work areas regularly using disinfestants. Another bacterial disease may also affect hostas, a shoot proliferation due to *Rhodococcus fascians*. Strict sanitation measures are the best way to avoid this problem.

Several viruses attack hostas. Symptoms of *Hosta virus X* (HVX) include mosaic, chlorosis, and necrosis on leaves. HVX can be transmitted very easily by hands or tools that come into contact with contaminated plant sap. *Impatiens necrotic spot virus* (INSV) causes discrete, circular, concentric rings, green to off-white, that look like a bull's-eye and may grow together when a leaf is heavily infected. Symptoms of *Tomato spotted wilt virus* (TSWV) are similar to those of INSV. In addition to being spread during vegetative propaga-

The fungus *Sclerotium rolfsii* often develops white mycelium on hosta leaf or stem tissue it has killed. The black structures at right are sclerotia, long-term survival structures that allow the fungus to persist in garden soil from season to season.

A *Phytophthora* species has caused a blighting of the foliage on this hosta.

Hosta petioles collapse when *Sclerotium rolfsii* attacks plants near the soil line.

tion, INSV and TSWV are transmitted by the western flower thrips. *Tomato ringspot virus* (ToRSV) produces chlorotic leaf spots with indefinite margins that fade from yellow to green. It is transmitted during vegetative propagation, by root-feeding nematodes, and possibly by pollen. *Tobacco rattle virus* (TRV) and *Arabis mosaic virus* (ArMV), which may be spread by propagation or by nematodes in the soil, may also cause chlorotic mottling and distortion on leaves of hostas. Only virus-free plants should be used for propagation. Carefully inspect all plants for virus symptoms before buying, and purchase certified virus-free plants. Using insecticides to control thrips may reduce transmission of INSV and TSWV in nursery production.

Two types of nematodes are pathogens of hosta. Foliar nematodes (*Aphelenchoides*) are microscopic worms that swim in films of water on leaves and stems, entering leaves though natural openings. They can move from plant to plant in drops of splashing water, on garden tools, or as plants are handled. Damage appears, on older leaves first, as stripes of light green to yellow leaf tissue parallel to the major veins. Affected leaf tissue turns brown and may drop out or tear, giving the leaves a tattered appearance. Carefully inspect plants for symptoms prior to introducing them into a nursery or landscape, and buy from reputable growers. Reducing overhead watering will help to prevent the spread of foliar nematodes, and removing damaged leaves also helps. Root knot nematodes (*Meloidogyne*) cause knots or small galls of root tissue, as well as branched root tips. Plants may appear stunted, wilt easily under stress, and may show signs of nutritional deficiency. It's important to inspect plants prior to planting, and to avoid replanting hosta in an infested area. For nurseries, rotation with a non-susceptible host, such as small grains, can reduce nematode populations in the soil.

Slug feeding commonly causes holes to appear in hosta leaves "overnight".

Deer are especially fond of hosta leaves.

Round, haloed leaf spots on hosta caused by an anthracnose fungus (*Colletotrichum* sp.)

Cercospora leaf spot on hosta.

A soft wet collapse of hosta leaves may be caused by bacterial soft rot (*Pectobacterium* sp.).

Impatiens necrotic spot virus (INSV) on hosta.

Tobacco ringspot virus (TRSV) on hosta.

Tomato ringspot virus (ToRSV) on hosta.

Tobacco rattle virus (TRV) on hosta.

Hosta virus X (HVX) on hosta.

Arabis mosaic virus (ArMV) on hosta.

Hosta virus X (HVX) on hosta.

Hyacinthus

Liliaceae

Hyacinthus, or hyacinth, is a familiar genus with large round bulbs, thick, strap-shaped leaves and a very showy, very fragrant early spring flower. The common hyacinth, *H. orientalis,* originated from the Mediterranean region. Bulbs should be mulched for winter protection in the northern United States. The planting area should have good drainage to avoid winter losses. Hyacinths may be forced for earlier bloom in shallowly-planted pots or in glasses of water.

The most common disease on hyacinths is yellow rot, caused by a species of the bacterium *Xanthomonas.* Yellow rot produces water-soaked streaks and yellowing on the leaves, followed by death of the flower stalks. Bulbs can sometimes be completely rotted, but you can cut across vascular bundles of less severely diseased bulbs to check for dark discoloration and bacterial ooze. Another bacterial disease is soft rot caused by *Pectobacterium*

Rot and basal decay of these hyacinth bulbs is due to *Fusarium oxysporum.*

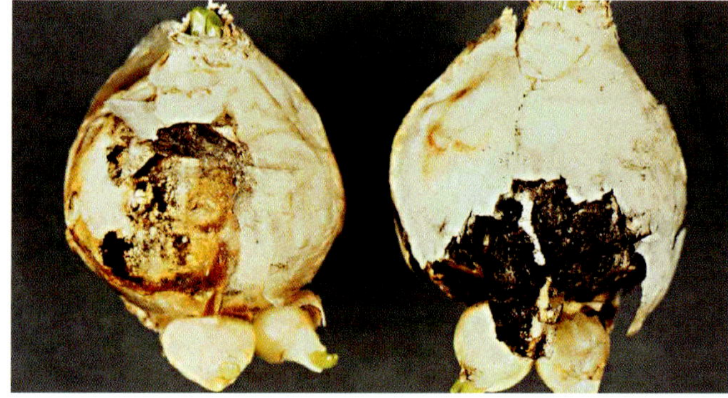

Dark black sporulation indicates infection by *Aspergillus niger* on these hyacinth bulbs.

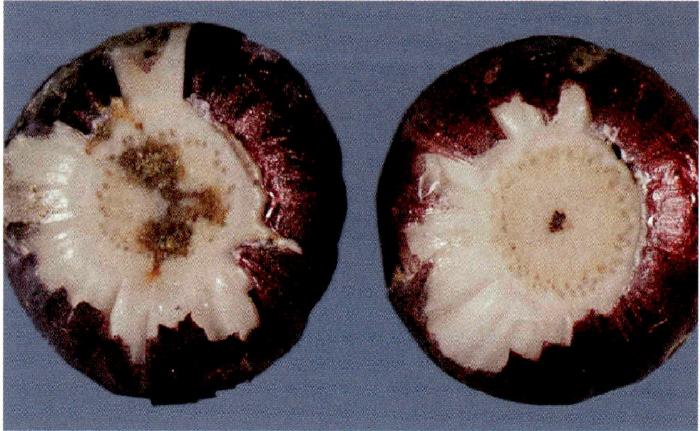

The discoloration in the hyacinth bulb at left is due to infection by bacteria (*Xanthomonas* sp.).

Bacterial soft rot (*Pectobacterium* sp.) in hyacinth bulbs.

carotovorum. Bulbs may be blind (fail to develop flowers) or flower expansion may be partial and flowers will rot. The leaves and flower stem may be easily pulled off of the bulb. Several kinds of stress will predispose bulbs to soft rot: excess heat during forcing, storage or shipment, or repeated freezing and thawing.

Botrytis hyacinthi may attack leaf tips and developing flowers. Sporulation of the fungus appears as a gray moldy coating of rotted areas. Rogue out any plants with Botrytis blight as soon as possible, to prevent the spread of the disease. Lumpy black sclerotia may be evident on bulbs attacked by either *B. hyacinthi* or *Sclerotinia bulborum*. Storage rots of hyacinth may be caused by species of the fungi *Aspergillus* and *Penicillium*. Leaves may show spots caused by the fungus *Stagonospora curtisii*, which can be carried on the dried scales of the bulbs. An anther smut caused by a *Ustilago* sp. fungus can result in failure to flower in extreme cases; this disease has been reported from Louisiana. Pythium root rot on hyacinths will cause them to be stunted, while the roots themselves will look glassy and eventually softened and gray. Sometimes tip dieback appears on leaves. The related organism *Phytophthora* sp. may also cause a bulb rot.

A ring of softened, discolored tissue in a hyacinth bulb is the classic symptom of stem and bulb nematode (*Ditylenchus dipsaci*). Remove and destroy all infected plants and plant debris. In nursery production fields and landscapes, rotating away from plant hosts of this nematode for 2 to 4 years is recommended.

Viruses, including *Hyacinth mosaic virus* (HyaMV), cause leaf mosaic and flowers on infected plants may fail to develop normally. Hyacinths infected with the phytoplasma disease aster yellows develop pale leaves, malformed flower parts and poor root systems.

A Penicillium leaf spot on hyacinth.

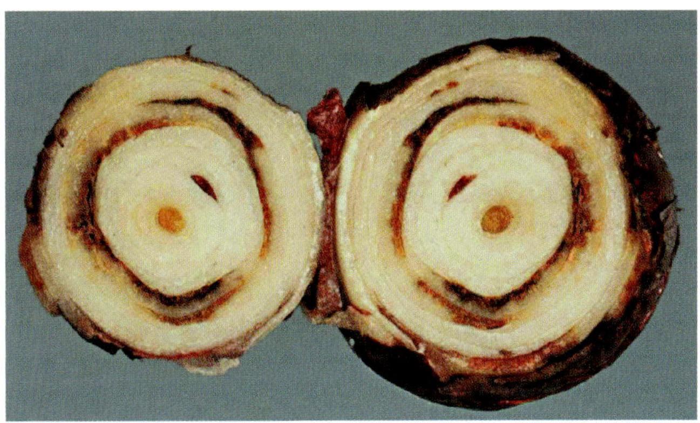

Stem and bulb nematode injury to hyacinth bulb.

Distortion of hyacinth flowers and leaf mosaic caused by hyacinth mosaic virus (HyaMV).

Hypericum

Hypericaceae

Hypericum, or St. John's-Wort, includes many species of shrubs, sub-shrubs and herbaceous perennials, some of which are native to the United States. These plants generally do well in moist, well-drained, humus-rich soil in part shade; dwarf species grow best in full sun in extremely well-drained soil suitable for rock gardens. Some are excellent ground covers. The invasiveness of *H. calycinum*, Aaron's beard, can be kept in check by shearing plants back every couple of years. Winter winds or hot summer conditions can desiccate foliage.

Hypericums include some very reliable garden plants, but they are vulnerable to several rust diseases caused by *Uromyces triquetrus* and *Melampsora* species. The rusts may cause chlorotic spots on the upper surface of the leaf, and pustules filled with yellow-orange sporulation, all of which form fungal leaf spots. In landscapes, rust symptoms can be severe if left unchecked. White spots or whitish coatings of fungus are due to powdery mildew. A bacterial (*Xanthomonas* sp.) leaf spot has also been observed. Avoiding excessive leaf wetness is important for minimizing rust and leaf spots, so overhead watering should be done from late morning to midday, thoroughly and less often, rather than doing frequent light waterings.

Root or crown rot problems have been noted from *Pythium* and *Phytophthora* water molds and by species of *Rhizoctonia, Fusarium* and *Thielaviopsis* fungi. To minimize the risk of these problems, it is helpful to avoid prolonged periods of waterlogged soils and minimize stressful growing conditions on the plants. In nursery fields and some landscapes, fungicide drench treatments may be helpful. Roots should be examined on plantings with declining vigor; this may be a sign of root knot nematode (*Meloidogyne*) infestation, which can be recognized by finding small swellings along the roots. Infected plants should be pulled up and destroyed; avoid replanting these sites with hosts of root knot nematodes for several years.

Tobacco mosaic virus (TMV) has been reported from St. John's-Wort. Infected plants should be removed and destroyed.

A rust fungus will cause angular yellow spots on leaves of St. John's-Wort; turning these leaves over will expose the colorful orange pustules of the fungus.

St. John's-Wort grown in a poorly drained site shows a blackened, stunted root system due to Pythium root rot.

Symptoms like those of root rot may be caused by root knot nematode, shown here on St. John's-Wort.

Angular leaf lesions due to a bacterial leaf spot (*Xanthomonas* sp.) on St. John's-Wort.

Iberis

Brassicaceae

Iberis, or candytuft, is primarily represented in perennial gardens by *I. sempervirens*, evergreen candytuft, which is a subshrub. Iberis is suited to full sun and well-drained soils. Liming the soil is beneficial, as neutral to alkaline soils are preferred. Iberis will tolerate soils of poor fertility. Cutting back after flowering every few years helps to rejuvenate plants.

If powdery mildew, gray mold blight (*Botrytis cinerea*), or Alternaria leaf spot (*Alternaria* sp.) appear, they can be suppressed by increasing spacing among plants, selecting sites with good air flow, and applying fungicides as needed. Downy mildew (*Hyaloperonospora parasitica*), white mold (*Albugo candida*), and the bacterial disease Xanthomonas leaf spot (*Xanthomonas campestris* pv. *campestris*) can be managed by keeping the foliage as dry as possible and improving air movement in the planting. Fungicides are also an option for downy mildew and white mold. Copper-containing fungicides can be used to suppress Xanthomonas leaf spot, but since many plants in this family are sensitive to copper products, they should be used with caution. Another bacterial problem, a shoot proliferation disease caused by *Rhodococcus fascians*, is found occasionally. It causes growth abnormalities in affected plants. Plants with stunted, distorted clumps of shoots at the base should be destroyed.

Alternaria leaf spots on candytuft are accompanied by yellowing.

Phoma root rot, caused by the fungus *Phoma lingam*, affects the roots primarily, but also the stem at the soil line. Control includes avoiding planting iberis where plants of the cabbage family (including cabbage, stock, and sweet alyssum) have been grown in the past several years, and keeping the foliage dry when watering. The fungus *Leptosphaeria* causes stem cankers, and results in areas of dieback within a planting. Other diseases that affect the roots or stem include Rhizoctonia damping off and stem rot (*Rhizoctonia solani*), Pythium root rot, and, in warmer regions of North America, Southern blight (*Sclerotium rolfsii*). Avoiding soil moisture extremes helps to minimize risk of Rhizoctonia problems, and staying away from waterlogged soils helps to avoid Pythium rot. Plants showing root or stem rot symptoms should be removed and destroyed. Fungicide drenches can also be used to suppress these soilborne diseases.

Iberis can also develop a disease known as clubroot, caused by the primitive, slime-mold-like organism *Plasmodiophora brassicae*. Symptoms include swollen and distorted roots, as well as wilting of the foliage. Clubroot is a risk primarily in poorly drained sites where water pools for long periods after rain or irrigation. To avoid clubroot, modify the site's drainage or move the plants to better-drained sites. Another pest that can result in distorted roots is root knot nematode (*Meloidogyne* sp.); this problem is likely to be more severe on sandy-textured soils. Infested plants should be removed and destroyed, and the site can be replanted with species that are not hosts of root knot nematode.

Aster yellows, caused by a phytoplasma, and several virus diseases have been noted on candytuft. Plants infected with phytoplasmas and viruses should be removed and destroyed to slow spread of these diseases.

Xanthomonas leaf spots on candytuft may coalesce into larger necrotic areas.

Downy mildew on candytuft may appear as yellowing or blighting of foliage. Close observers may see the whitish evidence of downy mildew on the infected leaves.

Pythium root rot has caused death of a portion of this candytuft plant.

Stem cankers caused by *Leptosphaeria* sp. on candytuft lead to yellowing, wilting and death of stems.

Growth of adventitious buds at the base of a candytuft plant caused by *Rhodococcus fascians* has led to development of a group of abnormal, stunted shoots.

Iris

Iridaceae

There are many species of *Iris* as well as many hybrid varieties, particularly for bearded iris. The acceptable conditions for the growth of most types of irises are a sunny or lightly shaded site and rich soils that are well drained. Some species, however, will grow happily with their root system submerged at the edge of a pond. The bearded irises are rhizomatous; this group includes Japanese iris (*Iris ensata*) and Siberian iris (*Iris sibirica*). Bulbous irises include Dutch iris (*Iris hollandica*), Spanish iris (*Iris xyphium*), and the dwarf *Iris reticulata*.

Bearded irises, *Iris germanica* hybrids, are grown in sunny or lightly-shaded settings and do best in soils that are neutral or slightly acidic, pH 5.5 to 6.5. Plant rhizomes horizontally with their upper surface exposed except in the hottest of climates, where they may be buried shallowly. Although these irises should be given thorough waterings during drought, and need moisture during the flowering period, note that heavy soils will predispose them to soft rot. The miniature dwarf bearded iris hybrids used in rock gardens are particularly dependent upon good drainage. Gathering leaf debris in fall is important for managing the iris borer, which overwinters in the leaves of tall bearded irises. This pest combines forces with bacteria in the genus *Pectobacterium* to kill plants. Divide every 3-4 years to alleviate overcrowding; else the centers of clumps will die out over time. Aphids and thrips can be troublesome; both can carry viruses to irises. Iris scorch is a malady of unknown cause in which leaves suddenly redden or brown as the plant dies.

Iris germanica, German Iris

Iris ensata, Japanese Iris

Iris sibirica, Siberian Iris

Iris cristata, Dwarf Crested Iris

Iris pseudacorus, Yellow Flag

Iris versicolor, Blue Flag

Iris ensata, Japanese iris, blooms later than many of the others. This type does well either standing in water or in rich, slightly acid soils with a high moisture holding capacity. Japanese iris will not thrive in dry gardens, and do not tolerate the addition of lime. Full sun or afternoon shade is acceptable.

Iris sibirica, Siberian iris, blooms largely in the interval between bearded iris and Japanese iris and is a good choice for low-maintenance gardens. Siberian iris perform best at a pH of 7.0 or below (slightly acidic), in soils rich in organic matter, in bright sun or light shade. Irrigate them thoroughly in drought periods unless they are grown in a continually boggy area. These are relatively trouble-free irises, not very prone to bacterial soft rot or iris borer. Divide clumps with minimum disturbance only after flower numbers have become noticeably reduced.

Iris cristata, crested iris, is suited to woodland conditions, as it thrives in part shade and will need supplemental irrigation in sunnier sites. It may also be grown in sun, and does well in moist soils with a pH near neutral that are rich in organic matter. This species, native to eastern North America, is often used as a groundcover.

Sclerotinia stem rot on iris (note the black sclerotia).

Iris pseudacorus, Yellow Flag, will form large patches in very moist areas, such as in or near lakes or streams. These plants are sometimes invasive in waterways. Yellow Flag can be grown in perennial borders with supplemental irrigation.

Iris versicolor, Blue Flag, is native to parts of the United States and Canada. It does well in wet areas, but can be grown in borders if the soil is humus-rich and water is supplied by rain or irrigation.

Bearded iris are especially prone to bacterial soft rot, which develops after entry of the bacterium *Pectobacterium carotovorum* through wounds. Wounds from leaf feeding by larvae of the iris borer are one of the main routes by which the pathogen gains entry to irises. *P. carotovorum* rots rhizomes and lower leaf sheaths. Plants can show either sudden wilt or slow dieback from the leaf tips. The rhizome and sheath rot is slimy and produces a foul smell. To reduce damage from soft rot, cut off and discard the rotted portions of rhizomes when making plant divisions, clean the remaining sections of rhizome, and let them dry in the sun before replanting in a new area. Plant shallowly, with the upper half of the rhizome above the soil surface. You can control the iris borer directly by squashing young larvae as they tunnel through leaves, and by cutting out large larvae from rhizomes when making divisions. During the late fall, remove and destroy old iris leaves, stems, and any nearby plant debris in order to prevent overwintering of iris borer eggs.

The important fungal leaf spots of iris are caused by *Heterosporium gracile* (sexual stage is *Mycosphaerella macrospora*, formerly known as *Didymellina macrospora*) and *Heterosporium iridis*. These start as small brown spots with water-soaked margins on leaves of bulbous and rhizomatous irises. After flowering, the spots enlarge and develop a gray center with a darker border of reddish brown. Masses of fungal spores can be seen in the center of the spots if you look closely. Brown streaks form as the spots run together,

Collapse of rhizomatous irises due to bacterial soft rot caused by *Pectobacterium* sp.

Bacterial soft rot in iris bulb caused by *Pectobacterium* sp., along with infection by *Sclerotinia bulborum.*

and entire leaves may wither and die, gradually weakening plants. Rhizome-forming irises are generally more susceptible than bulb-forming types. Removing and destroying old leaves in the late fall is important to keep the fungus from overwintering. Foliage can be protected against Heterosporium leaf spot with fungicide sprays beginning in early spring, adding a spreader-sticker to help the fungicide stick to leaves.

Ink spot is caused by the fungus *Bipolaris iridis* (formerly called *Mystrosporium adustum*). Leaf spots appear as dark, irregular blotches early in the spring. The foliage may be killed in wet weather. There are also black stains on the husks of bulbs that may form a ring pattern. Bulbs may be entirely rotted, leaving only black powder in a husk shell. Because the fungus may spread, this disease should be eradicated if it appears in the garden, by carefully removing the affected plants, including the bulbs and adjacent soil.

Bacterial leaf spot, caused by *Xanthomonas campestris* pv. *tardicrescens*, produces dark, water-soaked streaks on leaves that become especially apparent during humid weather. The water-soaked appearance of young bacterial leaf spots is similar to symptoms of Heterosporium leaf spot, minus the grayish masses of fungal spores in the centers of mature spots. Streaks caused by bacterial leaf spot can also be confused with injury caused by tunneling of the iris borer. Droplets of bacterial ooze will sometimes collect along the rim of spots. Sanitation practices (i.e. collecting and destroying leaves in winter, exposing the soil surface to sun, and avoiding splashing the leaves when watering) should help to curb future infections.

Rust on iris, caused by *Puccinia iridis*, is easy to identify from the brown powdery masses of spores that form in blister-like pustules on the leaves. Foliage may be killed in severe infestations. Removing leaves in the fall will help to destroy the overwintering fungus. Fungicides may be applied as protectants the following year.

If leaves are first injured by frost, the tips may be blackened by a weakly pathogenic fungus, *Didymellina*

Heterosporium leaf spot on iris causes individual oval spots that may coalesce into streaks or blotches.

Leaf spots and streaks caused by Heterosporium leaf spot on iris.

Early symptoms of iris rust.

Mature rust sporulation on iris.

Severe rust infection on iris may lead to leaf dieback.

poecilospora. The common name of this minor problem is black tip. The problem areas can simply be clipped off.

The soilborne fungus *Sclerotinia bulborum* causes black slime disease. Bulbs become blackened at base or tip, and should be culled out when planting. The fungus can be carried on the bulbs; bulb dips and dusts with fungicides have been used for counteracting this disease. Other species of *Sclerotinia* have also been reported from iris.

Iris bulbs infected with *Drechslera iridis* vs. healthy bulbs at right.

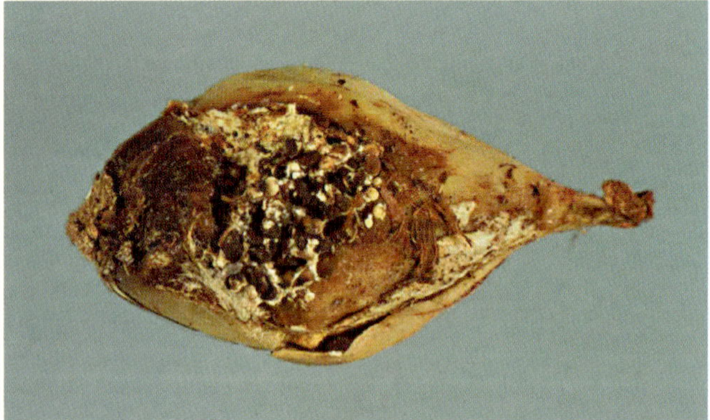

Iris bulb bearing sclerotia of the fungus *Sclerotium rolfsii* on the rotted area.

Iris bulb rot due to mixed infection of *Fusarium* and *Pythium* spp.

Iris bulbs may be attacked by various *Penicillium* species. These fungi have distinctive powdery, blue-green spore masses that develop on spotted areas of the bulbs. The bud scales become water-soaked at first, but eventually turn yellow or brown. Often a bruise or a wireworm feeding injury is the entry point for *Penicillium*. Infections at the base of bulbs cause the greatest harm. Control includes preventing bruises to the bulbs, curing them rapidly, and storing them properly to avoid overly warm temperatures or too much moisture.

In warmer parts of North America, rhizomatous irises showing tip dieback may be infected by the soilborne fungus *Sclerotium rolfsii* at the bases of the leaves. *S. rolfsii* forms white fan-shaped mats of mycelium as well as numerous spherical, white to reddish brown, pinhead-size sclerotia on the leaf bases. This same pathogen affects bulbous iris, causing plants to become chlorotic and stunted. Both the bulb and the base of the plant can be attacked. Bulbs may develop a soft white rot with a cheesy texture that will dry to show sunken tan areas. Sclerotia on the bulbs help to identify

Bulb rot due to *Fusarium oxysporum* on iris.

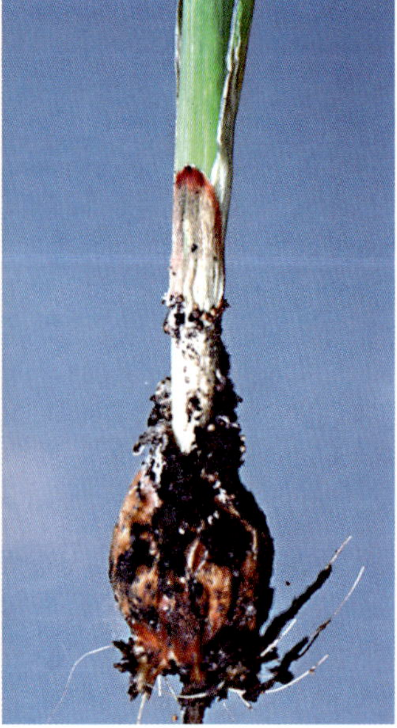

Iris stem base with white coating of *Sclerotium rolfsii* mycelium (note the tan sclerotia in the closeup).

the pathogen. Space irises rather than crowding them, and promptly remove diseased plants as soon as they are detected. Bulb soaks and soil fungicide drench treatments have been used for managing this disease. *Fusarium oxysporum* causes a basal rot of bulbous iris. Fungus growth on sunken areas of the bulb is whitish, sometimes with red pigmentation, but the bulb remains fairly firm. There is often a sharp margin between browned and healthy tissue. Care should be taken not to bruise the bulbs. Treating bulbs with a fungicide dip just after the harvest has worked well to reduce problems with this disease. Do not replant in an area that has become infested. The fungus *Rhizoctonia solani* causes a neck rot at the top of iris bulbs, killing leaves. Occasionally the bulbs are attacked directly, causing large dead brown areas. Soil fungicide drench treatments may be used in beds prior to planting. Rhizomatous iris can be killed during winter or spring by the fungus *Botryotinia convoluta*, which causes gray mold rot. Plants will die because their roots have shrivelled and decayed. Masses of spores appear on the rotted rhizomes in the spring, and large black sclerotia form on the lower part of decayed rhizomes. In the garden, plants that show noticeably weak growth in the spring should be checked for sporulation or sclerotia, and dug out and removed along with the adjacent soil if these are found.

The foliage of irises infected by a virus may be mottled with yellow, or show ring spots. On white, blue and lavender bulbous iris, dark teardrop-shaped marks may appear, and white feathery marks on flowers. Bud sheaths may also show blotches or streaks of discoloration. Plants with these symptoms should not be used for propagation, as the virus will be carried in the divisions. Virus-infected plants cannot be cured, and should be removed and destroyed to prevent spread by certain species of insects and nematodes. Several aphid-borne viruses have been reported from bulbous iris, including iris mild mosaic (IMMV), iris severe mosaic (ISMV), bean yellow mosaic (BYMV) and narcissus latent (NLV, previously called iris mild yellow mosaic virus). More recently, iris yellow spot (IYSV) has been identified on Dutch iris. Iris yellow spot is a tospovirus closely related to tomato spotted wilt and impatiens necrotic spot viruses: IYSV is largely a problem on onions, but iris and lisianthus are also hosts. Unlike TSWV and INSV, IYSV is spread by onion thrips (*Thrips tabaci*).

The cause of iris scorch, a progressive dieback from the leaf tips down to the rhizome, is uncertain. Roots also rot back to the rhizome. Abiotic (non-contagious) factors associated with the appearance of scorch include a combination of poorly drained soils and cold soil temperatures, high-temperature periods, and excess nitrogen fertilizer. Microorganisms that have sometime been found in association with scorch include *Pseudomonas* sp. bacteria and unidentified phytoplasmas. It is helpful to remove scorched plants from the garden; to save plants, the healthiest portions can be replanted in a new site.

The topple disease of Dutch iris, in which the top segment of the flower stem becomes water soaked and collapses along with attached flowers, is an abiotic problem. It is associated with calcium deficiency, and can be prevented by liming soils or treating with calcium nitrate fertilizer.

Chlorotic mottling and necrosis caused by an unknown virus on iris.

Closeup of symptoms of *Iris yellow spot virus* (IYSV) on onions; symptoms are quite similar on iris.

Yellow mottling caused by *Iris yellow spot virus* (IYSV) on iris.

Kniphofia

Liliaceae

Kniphofia, or torch lily, species and hybrids grow well in sunny gardens with perhaps some mid-afternoon shade. A very well-drained, fertile soil that is high in organic matter is desirable. If flowers appear sparse, this may indicate drought during bud development. Mulch for winter protection, or lift and store; divide as infrequently as possible. Leaves become unattractive after flowering and may be cut back.

Torch lily is generally disease-free. Rhizoctonia root and stem rot occasionally occurs. It is caused by the soilborne fungus *Rhizoctonia solani,* which enters the plant through the roots or attacks the stem at soil level. To minimize this problem, avoid moisture extremes. Diseased plants should be removed and destroyed. Labeled fungicides can be used if needed and can be very effective against this disease. Other reported problems on kniphofia include leaf spot caused by *Alternaria* and root galling due to the root knot nematode (*Meloidogyne*).

Dieback of kniphofia foliage due to Rhizoctonia crown rot.

Lamium

Lamiaceae

Lamium, or dead nettle, includes a few species that are popular ground covers. *Lamium maculatum*, the spotted nettle, is available in a number of varieties that provide a more compact groundcover in northern U.S. gardens. Plants may be sheared midsummer to keep them more compact. Even moisture, good drainage and some shade are desirable. Bare spots may develop in spotted nettle plantings that are exposed to drought, but they tolerate dry shade better than many other perennials. *Lamium galeobdolon*, yellow archangel, has in the past been known by the name *Lamiastrum*. It is a vigorous yellow-flowered ground cover that is grown over a wide range of lighting conditions depending on climate, with sunnier settings acceptable in the northern United States and deep shade acceptable in the South. Its invasiveness is related to soil fertility; plants are less invasive in poor soils, so slower growing varieties of the species should be chosen for more fertile ground. Slugs and snails may trouble lamium.

Diseases of lamium foliage include downy mildew. This fungus-like organism (*Peronospora lamii*) causes patches bounded by veins to show purple discoloration in some cases; sometimes they have a water-soaked appearance. On the underside of the leaf, a thin tuft of bluish-grey fungus can sometimes be seen under the discolored patches. Powdery mildew also occurs on *Lamium*. Sometimes you have to examine the leaf closely to see the white fungal threads on the surface of the leaf when the disease begins. Septoria leaf spot can be separated from the other common leaf diseases because the causal fungus forms spores in tiny, dark brown structures within the spots. Most leaf spots require wetness on the leaves in order for the pathogen to infect. Therefore, timing sprinkler irrigation so that foliage will dry as quickly as possible, combined with fostering good air circulation among plants (thinning plantings and removing leaves and other debris from among plants) tends to inhibit leaf spot fungi. Fungicides can protect healthy leaves from becoming infected if combined with keeping moisture off the leaves as much as possible.

Symptoms of foliar nematode (*Aphelenchoides*) start as light green areas on green-leaved culitvars, and more intense pink areas on pinkish-white leaved varieties. These areas, bounded by the veins, have an angular shape. As

Discolored patches in lamium leaves caused by downy mildew.

Interveinal browning due to foliar nematodes on lamium.

feeding continues, the angular areas die and turn brown. Tearing this tissue in a large, clear drop of water in a Petri dish will release these microscopic worms. They have a snake-like motion that can be seen with a good hand lens. Foliar nematode infected plants should be removed from the bed and destroyed. The nematode can spread to many other types of plants when the foliage is wet and leaves are touching one another or are being splashed by overhead irrigation.

Virus symptoms appear in lamium on occasion. Severe leaf distortion has been seen in the popular variety 'Beacon Silver' in North America.

Root and stem rot can be caused by at least three pathogens that make the stems shrivel, collapse, and have a dry appearance at the soil line. *Rhizoctonia solani* is the most common of these. *Sclerotinia* species usually produce a white mass of fungal threads at the base of the plant. In Phytophthora crown rot, a wet looking area develops near the soil line and the stem blackens as the plants collapse quickly and die. Fungicides are available for managing these problems in nurseries. Myrothecium crown rot may also occur (primarily during nursery production). Avoiding wounding can reduce disease but sometimes fungicides are needed to prevent losses from *Myrothecium roridum*.

The distortion on these lamium leaves is due to a caulimovirus infection.

Extensive dieback of a nursery-grown lamium plant with a Phytophthora crown rot.

Myrothecium crown rot on lamium.

Stem blackening caused by *Phytophthora citrophthora* on lamium.

Lavandula

Lamiaceae

Lavandula, or lavender, is famous as an herb and also important as a garden plant; these are shrubs or subshrubs rather than herbaceous perennials but are grown along with herbaceous perennials in nurseries and gardens. Common lavender is *L. angustifolia*; a hybrid of this plant with *L. latifolia* is called lavandin. Lavender varieties and hybrids demand good drainage and full sun, and do best in areas of the United States with a Mediterranean climate. High rainfall and high humidity do not suit them. Winter injury is common.

Leaves can be spotted or even killed by the fungus *Septoria lavandulae*. *Botrytis cinerea*, the fungus that causes gray mold, appears as fuzzy, gray growth on the affected leaves and stems. Both fungi require water on

A Botrytis canker is causing the symptoms on this lavender plant.

Xanthomonas leaf spot on lavender.

Phyllosticta blight on the lower foliage of lavender.

Fusarium wilt symptoms on lavender.

the surface of the plant in order to infect and both can be treated with fungicides to suppress their activity. Dark brown dead spots, limited in shape by the leaf veins, form on leaves infected with the bacterium *Xanthomonas campestris*. No chemicals effectively prevent this disease, and it can be especially problematic during the rooting process.

Phyllosticta blight results in the death of major branches of the plant, usually starting near the soil. Similarly, *Phytophthora* species can attack plants at the stem base when the weather is wet and warm, causing long, dark, slightly sunken areas on the stem. Plants eventually completely collapse and die when infected by *Phytophthora*. This disease causes sudden death of containerized lavender during the late summer and early fall. Preventive fungicide applications can be effective, but trying to stop losses once they start is usually fruitless. The fungus *Armillaria mellea*, commonly associated with the root and butt rot of woody plants, can also infect and kill lavender. *Macrophomina* may cause charcoal rot, easily identified by innumerable tiny black sclerotia that may be found on or under the bark or within the stems of lavender plants killed by the fungus. Wilt symptoms can be caused by species of *Fusarium* fungi. Generally, infected plants with root rots, stem rots, or wilts should be removed because fungicides are not particularly effective in protecting plants from these pathogens once symptoms are seen. Viruses are not usually a problem in lavender, but alfalfa mosaic (AMV) has been seen in this host in Italy, causing stunting, leaf deformity and chlorotic mottle, as well as poor flowering.

Wilting due to Phytophthora crown rot on lavender, compared to a healthy plant at right.

Round black bodies called sclerotia formed under lavender bark—the sign of charcoal root rot caused by *Macrophomina* sp.

Leucanthemum

Asteraceae

Leucanthemum ×*superbum* (*Chrysanthemum* ×*superbum*, also called *Chrysanthemum maximum*), Shasta daisy, is a dependable plant for sunny or lightly shaded locations in rich, well-drained soils. Plants perform well in the southern United States but tend to be short-lived there. In the western states they can be long-lived perennials returning year after year. Deadheading can stimulate additional blooms. Divide clumps every 2-3 years to maintain vigorous growth.

Alternaria leaf spot on Shasta daisy is a distinctive dark brown, almost black spot that often forms at the edge of the leaf in a target-like pattern, caused when the fungus *Alternaria* sp. forms spores on the surface of the leaf on a day/night cycle. The fungus *Septoria* sp. also causes a very black spot on the leaves. It forms spores in numerous tiny dots in the dead area. Both fungal leaf spots are severe during wet seasons or if sprinkler irrigation is frequent. Other fungal leaf spots are caused by *Cercospora*, *Botrytis* and *Cylindrosporium* species. These diseases tend to be more common during production or in climates that receive summer rainfall routinely. At the end of the growing season, as much of the above-ground plant debris as possible should be removed from the area because these fungi survive from year to year in plant debris. Insect injury can easily be mistaken for a fungal leaf spot on Shasta daisy: numerous yellow and brown spots on the leaf that appear to be slightly collapsed may be caused by plant bug feeding. Observe the plants with such spots carefully and watch for plant bugs — they will move rapidly to the underside of the leaf when you approach the plant.

Fusarium and Rhizoctonia stem rots begin at the crown of the plant, causing branches to wilt, turn dark brown, and die. If entire plants are killed, also look for a white mass of fungal growth at the stem base formed by *Sclerotinia* species or, in southern states, a white fungal growth followed by tiny white to golden-brown balls (sclerotia) formed by the fungus *Sclerotium rolfsii*. Rhizoctonia, Sclerotinia and Sclerotium stem rots are most severe in moist climates and where mulch or soil is piled around the base of the plants. All of these fungi survive well in the soil and in plant debris from year to year. Fungicides are available to suppress *Sclerotinia*, *Sclerotium* and *Rhizoctonia* but these are more useful during nursery production than in the garden.

Dark brown to black roots can be caused by the fungi *Thielaviopsis basicola* and *Fusarium* sp. or the water mold *Pythium* sp. All may be found in local soil. Fungicides can be applied to suppress root rots.

Alternaria leaf spot on Shasta daisy.

Botrytis leaf spot on Shasta daisy.

Wilting of Shasta daisy due to Rhizoctonia crown rot.

Cauliflower-like growths of green tissue close to the soil are caused by the bacterium *Rhodococcus fascians*. This shoot proliferation disease is propagated along with plants grown from cuttings taken from an infected mother plant. When flowers do not form properly and flower parts appear more like modified leaves, Shasta daisies may be infected with the aster yellows phytoplasma, which is carried from plant to plant by leafhoppers. Infected plants should be destroyed, as well as nearby weeds that can serve as aster yellows reservoirs.

Stunting and scorching of Shasta daisy caused by Thielaviopsis root rot.

Shasta daisy showing stunted shoot proliferation at the base of the plant caused by *Rhodococcus fascians*.

Lewisia

Portulacaceae

Lewisia is native to the western United States, primarily represented by *L. cotyledon* in American gardens. Lewisia is highly popular in the Pacific Northwest. Plants will do best in very well drained soils such as those used in rock gardens. Deciduous lewisias should be grown in full sun locations and evergreen types in light shade.

Small, yellow-orange masses of spores form in spots on the undersides of leaves infected by several rust fungi in the genus *Uromyces*. This fungus survives on live plants and in the recently killed leaves.

When *Rhizoctonia solani* attacks plants at the crowns, the entire plant browns and dies. This fungus survives well in soil and in plant debris; therefore, the latter should be removed from the bed at the end of the season. A fungicide drench can be applied the following spring to protect the emerging plants. Lewisias are also susceptible to various *Phytophthora* species that are commonly found in nurseries, so avoid overwatering them.

Liatris

Asteraceae

Liatris, or blazing star, is an American native used as a cut flower as well as a garden plant. Place it in full sun, in well-drained soil of moderate fertility. *L. spicata*, spike gayfeather, is the most widely planted species. Wet, heavy soil during the winter can be very damaging to *Liatris*. The foliage tends to deteriorate after flowering.

Several different fungi attack the leaves of liatris including *Phyllosticta*, *Septoria, Coleosporium* (a rust with pitch pine, *Pinus rigida*, as the alternate host), *Puccinia* (a rust with grasses as alternate hosts), and powdery mildew species. To suppress these fungi, avoid sprinkler irrigation, improve air circulation among the plants, and apply fungicides, if needed.

Rhizoctonia stem rots begin at the base of the plant, causing leaves to turn dark brown and die. *Sclerotinia* spp. can also kill the entire plant. Look for small white masses of fungal growth on the infected tissue formed by *Sclerotinia* species. If this disease is a problem be sure to check new corms before planting for signs of mold, sclerotia and rot. Wilting and death of plants may also be caused by *Verticillium albo-atrum*. All three of these fungi survive well in the soil and in plant debris. Removing plant debris at the end of the growing season and applying a fungicide drench

The stunted liatris at left have Verticillium wilt.

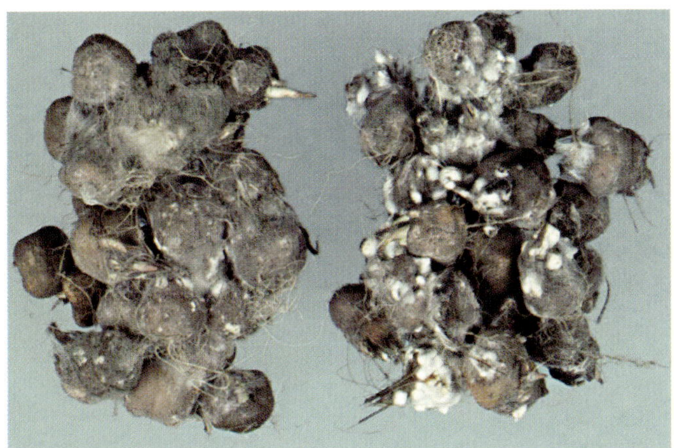

Sporulation of *Botrytis cinerea* (at left) and sclerotia of *Sclerotinia sclerotiorum* (at right) on liatris corms.

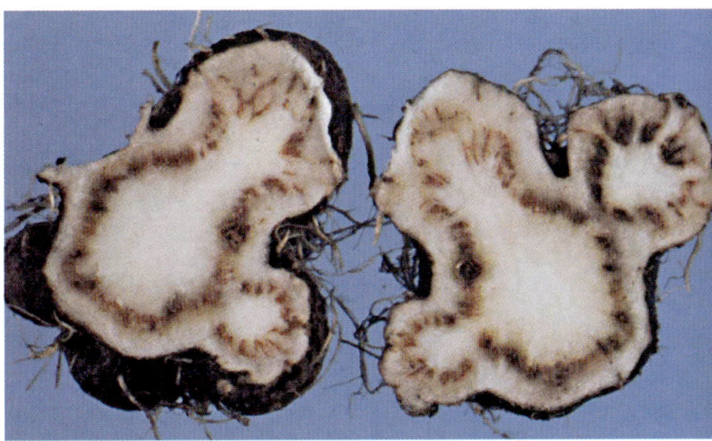

Internal vascular discoloration in liatris corms due to Verticillium wilt.

in the spring can control *Rhizoctonia* and *Sclerotinia* but not *Verticillium*. Pythium root rot, in contrast, is regulated primarily by soil conditions. To keep *Pythium* in check, avoid overwatering and poorly drained planting sites.

The stem and bulb nematode, *Ditylenchus* sp., may attack liatris, causing a distinctive reddish ring in the corms. Dispose of infected plants.

Rings of reddish discoloration in liatris corms are an indication of stem and bulb nematode infestation.

Liatris foliage coated by a powdery mildew.

Septoria leaf spot on liatris.

Discolored roots caused by Pythium root rot on liatris.

Sclerotinia stem rot causing collapse of liatris.

Lilium

Liliaceae

Lilium, or lily, includes both familiar and unusual plants in a wide range of forms and heights, some of which are used in full sun and some in partly-shaded locations. Native and exotic species and hybrids are grown, including the widely popular Asiatic and Oriental hybrids. Most lilies grow well in slightly acid soils, but there are some exceptions. The Martagon hybrids, including *Lilium* x *superbum*, like a pH of 6.5 to 7.5 and are less tolerant of acid soils than Asiatics and Orientals. Lilies generally prosper in soils that are well drained but high in organic matter. Excess moisture during the winter is very harmful. To keep lilies vigorous, remove spent flowers but keep stems on plants until they die back in the fall.

The fungus *Botrytis elliptica* is a common and aggressive leaf spot pathogen on lilies. Leaves develop brown, oval spots; if environmental conditions are favorable, infected leaves will turn brown and dry, beginning with the lower leaves and progressing upward. Maintaining good air circulation among plants and applying a fungicide can greatly reduce damage caused by Botrytis blight. The fungi *Botrytis cinerea, Cercospora*, and *Aureobasidium* can also cause leaf spotting. Foliar nematode (*Aphelenchoides* spp.) can cause brown streaks on leaves; avoid splashing water and crowding plants to keep these nematodes in check.

Lily mosaic (LMV), transmitted from plant to plant by aphids, causes yellow streaking or mosaic on the leaves as well as leaf distortion. Tulips often carry LMV and may provide a source of the virus for lilies in the vicinity. Tiger lily, *L. tigrinum*, is symptomless when infected with LMV, and therefore may also act as a source of virus for other lilies that will show disease symptoms. Several other viruses can occur in lilies, including *Cucumber mosaic virus* (CMV). These viruses may cause patches of light and dark green on the leaves, stunting of plants, or abnormally shaped leaves and stems.

Fusarium oxysporum causes basal rot, destroys the bulb and causes the foliage to yellow, especially in regions with hot summers. Root rot can be caused by *Pythium* and *Phytophthora* species, including *Phytophthora nicotianae*. Bulbs and stems are attacked by *Rhizopus*, *Sclerotium*, *Penicillium*, *Sclerotium*, *Rhizoctonia*, *Cylindrocarpon* and *Fusarium* species. For bulb disease prevention, bulbs need to be treated with fungicide before or at planting. Once symptoms of a bulb or stem rot are seen on lilies in a bed, it is usually too late to apply a fungicide to control the disease. If symptoms are seen in established plants, the infected plants should be collected and destroyed to reduce the chance of spread within the planting.

Mixed infection of lily bulbs with *Fusarium* and *Cylindrocarpon* species.

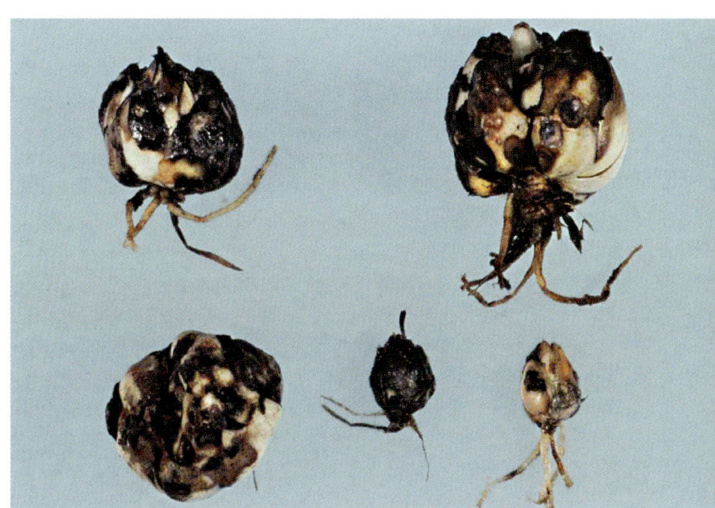

Lily bulb infection with *Sclerotium wakkeri*.

Botrytis blight on lily foliage.

Botrytis blight on lily flowers.

Botrytis leaf spot on Asiatic lily.

Pythium root rot on container-grown lilies.

Foliar nematode infection on lily.

Symptoms of *Lily mottle virus* (LiMoV) on lily.

Unknown virus on lily foliage showing mottling.

Limonium

Plumbaginaceae

Limonium, or statice, includes both annuals and perennials. *L. latifolium*, sea lavender, *Goniolimon tataricum*, German statice, and *L. perezii*, Perez statice, are some of the best-known perennials. Of these, Perez statice is the least cold-hardy. These are full sun plants in the northern United States but require some shade in the South. Well-drained, slightly acid soils are desirable for most forms of statice. These plants are suitable for sandy areas and many are drought and salt-tolerant. Long tap roots make it very difficult to divide statice plants, so they should be left undisturbed.

Statice is susceptible to a number of fungi that cause leaf and stem spots, including *Alternaria*, *Botrytis*, *Aschochyta*, *Fusicladium*, *Phyllosticta*, *Colletotrichum*, and *Cercospora*. It is also susceptible to rusts (*Uromyces limonii* and *Uromyces limonii-caroliniani)* that do not require an alternate host plant. Alternaria leaf spot is usually a purplish-brown circular spot on the older foliage while Botrytis-caused spots tend to be brown to tan in color. Cercospora leaf spot on some perennial statice appears as half-moon brown spots on the leaf-like tissue along the flowering stems. Some of the affected tissue has a wet appearance, at least at first. *Phyllosticta* sp. on stems causes a tan to gray dead area to form, and the fungus produces tiny, black, dot-like fruiting structures within the dead area. Downy mildew (*Peronospora statices*) produces bluish-gray fungal growth on the undersides of leaves and may severely stunt or kill plants, while powdery mildew appears as the typical white, dry, fungal growth on the upper surface of leaves. Foliar nematodes also cause a leaf spotting, typically appearing as angular dead areas between, and limited in shape by, the leaf veins. *Botrytis*, *Alternaria*, *Colletotrichum*, and *Cercospora* also blight the flowers, causing them to cease development or killing them after they have opened. All of these are favored by wet conditions, including conditions fostered by frequent use of sprinklers for irrigation.

Abnormally colored or shaped leaves may be caused by several different viruses, including clover yellow vein (ClYVV), tobacco rattle (TRV), tomato spotted wilt (TSWV), impatiens necrotic spot (INSV), and turnip mosaic viruses (TuMV). Infected plants should be removed and destroyed. Infected plants should not be divided to start new plants even if some sections appear normal. Generally,

Alternaria leaf spot on statice.

Downy mildew on statice shows bluish-gray sporulation on leaf undersides as well as leaf distortion.

the viruses infect all plant parts even if symptoms are not obvious. Note that insect feeding may also distort or discolor foliage. Aster yellows can cause flattening of the stems, distortion of leaves, and formation of leaf-like parts where flowers should be. As is the case with viruses, aster yellows infected plants should be destroyed.

Root and stem rots caused by *Phytophthora*, *Pythium*, *Rhizoctonia*, and *Sclerotium* kill portions of plants or entire plants. These pathogens may be found in the soil or in debris left from plants infected the previous year. Removing diseased plants as they are noticed and applying fungicide drenches in production areas are the primary control measures.

Colletotrichum sp. causes anthracnose on leaves and here is shown on flowers of statice.

Flower spots on statice caused by a *Cercospora* species.

Foliar nematode infestation on statice showing interveinal necrosis.

Powdery mildew on statice appears in white patches on upper sides of leaves.

Leaf distortion caused by infection of statice by *Grapevine Algerian latent virus* (GALV).

Liriope

Liliaceae

Liriope is a genus of evergreen plants often used as a ground cover or along walkways as an edging. The species *L. muscari*, blue lily turf, is especially reliable under conditions of drought, high temperature and humidity. It has become a staple in southern and western landscapes of the United States. These plants can be grown in full sun or heavy shade, but will fill in areas more quickly if provided with full sun. *L. spicata* is more shade tolerant than *L. muscari*. Slightly acid soils are preferred by liriope. Protect these plants from desiccating winds in regions with cold winters by using mulches. *Liriope* spp. may be cut back in early spring to renew the foliage.

Liriope anthracnose appears as tan spots surrounded by a dark brown zone, creating a target-like pattern. Within dark dots on these spots are the spore-forming structures of

Phytophthora leaf and crown rot on lily turf can be problematic in landscapes.

Anthracnose on lily turf can cause leaf spots.

Phytophthora leaf and crown rot on lily turf in the nursery.

the anthracnose fungus, *Colletotrichum* species. With a good magnifying glass, fringes of brown to black spines can be seen sticking up from the fruiting structures, sometimes accompanied by pale orange spore masses. Other leaf spots are caused by *Alternaria, Sphaeropsis* and *Cercospora* species. All of these will be spread when spores are splashed about during rain or overhead irrigation. Water in late morning to midday so that foliage can dry before nightfall.

Phytophthora, *Rhizoctonia*, *Pythium*, *Fusarium*, and *Sclerotium* can cause root rots or crown rots on liriope. Crown rots due to *Fusarium*, *Phytophthora* or *Pythium splendens* are particularly common. Phytophthora leaf and crown rot is caused by *Phytophthora palmivora*; this disease occurs across the Southeast, where the variety 'Evergreen Giant' has been found to be especially susceptible. The first sign of infection is yellowing at the base of individual leaves, progressing upwards. Young, immature leaves at the center of the plant turn brown and become water-soaked. Yellowing is evident on the older infected leaves. Root systems may also be rotted. Plants with symptoms of root or crown rot should be removed from gardens or nurseries. Excessive irrigation will increase problems with diseases caused by *Pythium*, *Phytophthora* and *Fusarium*.

Extensive crown rot on lily turf caused by *Fusarium* sp.

Close-up of Fusarium root rot on lily turf.

Lobelia

Campanulaceae

Lobelia includes species and hybrids of herbaceous perennials that do well in moist, humus-rich soils. Full sun or partial shade is suitable, with some late afternoon shade a requirement in the South and Midwest. A light mulch layer over lobelias in winter is helpful. Plants should be divided every 2-3 years.

Small dead spots on leaves can be caused by the fungi *Cercospora*, *Colletotrichum*, *Phyllosticta*, and *Septoria*. Lobelia is also susceptible to a smut (*Entyloma* sp.) and a rust (*Puccinia* sp.). Fungicides are available to protect new foliage against most of these fungi. When irrigating, it is helpful to reduce the time that leaves remain wet.

Lobelia is susceptible to several viruses that cause mottled light and dark green areas to form on infected leaves, including alfalfa mosaic (AMV), tomato spotted wilt (TSWV), impatiens necrotic spot (INSV) and beet curly top (BCTV). *Cucumber mosaic virus* (CMV) causes puckering of leaves, and sometimes mosaic symptoms. Aster yellows, caused by a phytoplasma, is also found occasionally on lobelia. In the past few years, quite a number of herbaceous perennials propagated from cuttings have been found infected with bacteria, including *Agrobacterium tumefaciens*, the agent of crown gall. Be sure to check all new cuttings and small plants for signs of galls or other abnormal growth at their bases before purchasing them or propagating them.

Root and stem rot can be the result of infections with *Rhizoctonia*, *Sclerotium*, *Sclerotinia*, *Phymatotrichopsis*, or *Pythium* species.

Propagation of lobelia by cuttings can spread crown gall.

White mycelium and lumpy sclerotia of *Sclerotinia* sp. on the root crown of lobelias with stem rot.

Tomato spotted wilt virus (TSWV) on lobelia appears as yellow mottling and slight distortion.

The aster yellows phytoplasma causes a condition called virescence (greening) of flowers as shown here on lobelia.

Cucumber mosaic virus (CMV) on lobelia can appear as yellowing, purpling or reddening on both leaves and stems.

Lupinus

Fabaceae

Lupinus, or lupine, includes only a few garden plants. The most commonly seen are the Russell hybrids. North American native species such as *L. polyphyllus* and *L. perennis* tend to be tolerant of poor, sandy soils. For the hybrid lupines, however, a rich, well-drained, neutral to acid soil is preferred. Because the plants are not tolerant of hot summers, lupine hybrids may be transplanted into southern gardens in the fall for spring bloom. Grow lupines in full sun or partial shade. Thorough, deep waterings and mulching are helpful in summer. Winter mulch may be helpful in colder climates. Remove spent flowers to maintain plant vigor. Plants are fairly short-lived; replant in the fall for best results.

Lupines are susceptible to a number of different leaf spots including those caused by the fungi *Alternaria*, *Ascochyta*, *Phyllosticta*, *Septoria*, *Stemphylium*, *Cercospora* and *Colletotrichum* species. In particular, anthracnose caused by *Colletotrichum* sp. can cause individual spots on leaves and continue to kill entire branches. Young seedlings as well as mature plants may be attacked, as this is a seed-borne disease. Spots on the cotyledons are often the first sign that anthracnose is present in a seed lot. All of the leaf-spotting fungi on lupines are favored by wet conditions and survive from year to year on plant parts that were infected the previous season. Fungicides are available to protect new foliage from attack.

Powdery mildew typically shows white fungal growth on the upper surface of leaves whereas downy mildew (*Peronospora*) has a bluish-gray growth of sporangiophores on the lower side of infected leaves. Two rusts, *Uromyces* and *Puccinia* species, form yellowish-brown spore masses on the undersides of leaves. Flowers, flower stems, and adjacent leaves can be blighted under wet conditions by *Botrytis cinerea*. Foliar nematodes (*Aphelenchoides*) feeding in areas between veins can cause death of that tissue, creating angular brown spots. The viruses bean yellow mosaic (BYMV), bidens mottle (BiMoV) and tomato spotted wilt (TSWV) are known to infect lupines.

Root and stem rots of lupines are attributed to *Fusarium*, *Macrophomina*, *Pythium*, *Phytophthora*, *Rhizoctonia*, *Thielaviopsis*, *Sclerotinia* and *Sclerotium* species.

The most common disease on lupine is anthracnose.

Anthracnose lesions often form half-moon shapes at the edges of lupine leaves.

In some cases, anthracnose on lupine appears as sunken cankers on stems.

Lesion (*Pratylenchus*), sting (*Belonolaimus*), ring (*Criconemella*) and root knot nematodes (*Meloidogyne*) feed on lupine roots. The small galls caused by root knot can be confused with those that result from the beneficial association between the lupine roots and the nitrogen-fixing bacteria, *Rhizobium*. *Rhizobium* sp. has similar beneficial associations with other plants in the legume family.

The beneficial bacterium, *Rhizobium* sp., causes galls to form on roots of host plants like lupine.

Leaf discoloration and distortion of lupine caused by *Impatiens necrotic spot virus* (INSV).

Stemphylium leaf spot on lupines occurs rarely.

Plant bug injury on lupine can be confused with fungal leaf spots.

Powdery mildew on lupine shows typical white patches on the upper surface of leaves.

Lychnis

Caryophyllaceae

Lychnis, or campion, belongs to the carnation family. These plants do well in light soil with good drainage. The lower foliage of *L. chalcedonica*, Maltese cross, may deteriorate during summer dry spells. *L. coronaria*, rose campion, in contrast, performs well in poor soils and dry conditions. Grow *Lychnis* spp. in well-drained soils in full sun to part shade, and provide some afternoon shade in southern gardens. Lychnis tend to be short-lived, but in many cases will reseed themselves in the garden.

Leaf spotting on lychnis can be the result of infection by *Alternaria*, *Septoria*, *Phyllosticta*, or *Leptothyrium* fungi. Phyllosticta leaf spots have a distinctive yellow halo. Other fungi that attack the leaves or flowers of lychnis include the rusts *Puccinia* and *Uromyces*. On flowers, there may be an anther smut fungus, *Ustilago* species. Flowers can also be killed by the fungus *Botrytis cinerea* under wet conditions.

Root and stem rots on lychnis may be caused by *Phytophthora* as well as the fungi *Phymatotrichopsis*, *Rhizoctonia*, *Corticium*, and *Sclerotium* species. Avoid overwatering, or planting in poorly drained soils, to prevent root and stem rot on lychnis. Keep mulch away from the stem to discourage Rhizoctonia stem rot.

Lysimachia

Primulaceae

Lysimachia, or loosestrife, does well in moist, rich garden soil. Some species are used as specimens and some, such as *L. nummularia*, as groundcovers. The native *L. ciliata* (syn. *Steironema ciliatum*), fringed loosestrife, and *L. punctata*, yellow loosestrife, are relatively drought tolerant. Loosestrifes are aggressive growers and some, including the gooseneck loosestrife, *L. clethroides*, can be invasive.

Leaf diseases may be caused by several fungi, including *Rhizoctonia*, *Cercospora*, *Phyllosticta* and *Septoria,* as well as the rust fungi *Puccinia* and *Uromyces*. Should foliar diseases become a problem, be careful to irrigate early in the day so that the foliage dries out swiftly. *Impatiens necrotic spot virus* (INSV) has been reported, with symptoms including an indistinct mottling. Management of the vector of INSV, the

Impatiens necrotic spot virus (INSV) on loosestrife can appear as yellow or black ringspots with distortion.

western flower thrips, is especially important during greenhouse propagation. Tobacco mosaic (TMV) may also infect loosestrife, and this virus is easily spread by handling plants. *Alfalfa mosaic virus* (AMV) is seen occasionally.

The root knot (*Meloidogyne*) and stem (*Belonolaimus*) nematodes cause galls and dead lesions on roots. Stem rots may be caused by fungi: *Sclerotium rolfsii,* which causes Southern blight in southern states, and *Sclerotinia* species are two of the most important stem pathogens on loosestrife.

Lythrum

Lythraceae

Lythrum, or loosestrife, is notorious for its invasiveness and competition with native plants. Sterile varieties of lythrum have been produced and widely planted in gardens, but research has now shown that these plants can still produce viable seed if grown near naturalized populations of *Lythrum salicaria* (a European native) that supply pollen. Thus, all lythrums should be cut back after flowering to prevent seed set. They grow best in full sun in moist soils. Because of their invasiveness, however, it is most prudent to avoid planting loosestrife.

The two most important diseases of lythrum are Septoria leaf spot and Rhizoctonia root rot. During production both can be problems since overhead irrigation is commonly employed. Try to reduce the time leaves are wet to minimize both diseases. Stem cankering due to a *Coniothyrium* sp. has also been observed on lythrum. One virus known to cause mottling, but which can also be in plants exhibiting essentially no symptoms, is *Cucumber mosaic virus* (CMV).

Brown stem canker caused by a *Coniothyrium* sp. on loosestrife.

Septoria leaf spot on loosestrife occurs in plant centers, where the leaves stay wet.

Maianthemum

Liliaceae

A few species of *Maianthemum*, or mayflower, may be used as groundcovers in shady areas. Mayflowers are ideal for moist, shaded sites. They need humus-rich, neutral to acid soil. *Maianthemum racemosum* (formerly named *Smilacina racemosa)*, or false Solomon's seal, is a woodland native. Cultivated forms like shaded, moist conditions and cooler climates. Acidic soil high in organic matter is desirable.

Fungal leaf spots are seen rarely, and are reported primarily from plants in the wild. These may be caused by *Cercosporella idahoensis, Cylindrosporium smilacinae, Heterosporium asperatum, Phleosporium vagnerae, Phyllosticta* sp., *Ramularia smilacinae, Septoria smilacinae, Sphaeropsis cruenta* and other fungi. The rounded, tan to brown spots on leaves caused by these fungi can occasionally cause leaves to turn yellow and wither if infection is severe. Remove and destroy foliage with severe symptoms. Leaf smut can be caused by the fungus *Urocystis colchici.* Control measures for leaf smut are similar to those for rusts, which may be caused by *Uromyces* or *Puccinia* species on this host. Fungicide use is an option in plant production environments for most foliar diseases. Be sure to get an accurate diagnosis, as a single fungicide will not control all leaf pathogens. Keep overhead irrigation to a minimum to avoid favoring fungal leaf spots, rust or leaf smut.

Malva

Malvaceae

Malva, or mallow, is a group of plants for well-drained, sunny sites in moist soils with moderate to high pH. Some are tolerant of partial shade. Hollyhock mallow, *M. alcea*, is drought tolerant. Mallows are relatively disease and insect prone. Weedy members of this genus may serve as hosts of hollyhock rust. The biennial *Alcea rosea,* or hollyhock, is often sold with perennials.

The most important leaf diseases on many species of *Malva* are rusts, caused by *Puccinia malvacearum* or *P. heterospora.* These fungi attack *Alcea*, *Sida*, *Hibiscus*,

Abutilon, and other species in the mallow family. Other leaf spots are caused by the fungi *Alternaria*, *Cercospora*, and *Septoria* spp. and bacteria in the genus *Xanthomonas*. All of these require wet leaves for infection to occur—thus, frequent rains or daily sprinkler irrigation favor their development. Mottling of leaves, stunting of plants, and leaf distortion may result when plants are infected with tomato spotted wilt (TSWV), malva vein clearing (MVCV), or beet curly top (BCTV) viruses.

Roots of mallows can be killed by species of the fungi *Rhizoctonia*, *Macrophomina* or *Phymatotrichopsis*.

The most common disease of mallow is rust.

Under production in the southern United States, mallow can be infected with a Xanthomonas leaf spot.

Mentha

Lamiaceae

Mints (*Mentha* spp.) are widely grown for culinary and ornamental purposes. They are popular ground covers that come in a wide variety of colors, with various aromas. They thrive in moist soils.

Tan to brown leaf spots may be caused by *Cercospora*, *Phyllosticta*, or *Septoria* species. Each of these fungi produce spores in small, dark fruiting structures that pepper the dead area of the leaf spot. Rust (*Puccinia* spp.) infection is seen as yellow spotting of the upper surface of the leaf. On the upper or lower leaf surface, orange-brown dusty spores form, sometimes in rings or target-shaped patterns. Powdery mildew has its typical appearance of white fungal growth on the upper surface of an infected mint leaf. Foliar nematode infection results in yellow, and later brown, angular areas, limited in size and shape by the leaf veins. Light and dark green mottling of the leaves can be caused by tomato ringspot (ToRSV) or tomato spotted wilt (TSWV) viruses. *Strawberry latent ringspot virus* (SLRSV) can also affect mints.

Wilting and death of mint species can result from infection by *Verticillium* species. These soil-borne fungi enter the roots of the plant and plug the water-conducting tissue (xylem). Frequently a one-sided wilt develops where branches on one side of the plant exhibit symptoms while the remainder of the plant appears normal. Eventually, the entire plant wilts and dies. Replace Verticillium wilt-affected plantings with species less susceptible to the disease. Stems can be infected by a number of different fungi including *Fusarium*, *Alternaria*, and *Phoma* species. When plants are infected with root knot nematode

(*Meloidogyne*), small swellings or galls form along the roots. Examine roots of plants before purchase to avoid bringing root knot nematode into the garden or nursery. Plants with root knot nematode should be discarded.

Mints are often infected with powdery mildew during production as well as landscape use.

Rust on many species and varieties of mint can be found in landscapes and nurseries.

Mertensia

Boraginaceae

Mertensia, or bluebells, includes a number of native species. The best known is *Mertensia virginica*, Virginia bluebells. These plants go dormant in summer, so their leaves will naturally turn yellow as they go into decline. Bluebells are grown in moist areas, in acid soil high in organic matter, and need some shade. They will spread well on their own if not disturbed, so division is not necessary.

While powdery mildew exhibits white fungal growth on the upper surface of the infected leaf, downy mildew (*Peronospora myosotidis*) forms bluish-gray tufts of fruiting structures on the undersides of leaves with yellow areas visible on the upper side. Tan spots that contain dot-like fungal fruiting structures (pycnidia) are caused by *Septoria poseyi*. Fungicides are available to protect healthy foliage from attack by powdery mildew and the various leaf spotting fungi. Two different species of rust, both in the genus *Puccinia,* attack bluebells, while a leaf smut is caused by *Entyloma serotinum*. Leaf mottling and yellowing may be caused by *Cucumber mosaic virus* (CMV), which is usually spread from plant to plant by aphid feeding.

Fungal stem rot can be caused by *Sclerotinia* or *Sclerotium* species. *Sclerotinia* produces a white mold growing in clumps along the lower stem, sometimes accompanied by lumpy black sclerotia. Sclerotium stem rot often includes a white, fan-like growth of fungus on lower stems and later small tan, ball-like knots of fungus (sclerotia), resembling mustard seeds. Both pathogens survive in plant debris or soil in sclerotia, which have been described as a 'fungus in a box with its lunch.' The fungi remain dormant within the sclerotia with a good supply of nutrients, awaiting favorable weather conditions and the presence of susceptible plants before they resume activity.

Monarda

Lamiaceae

Monarda, known as Oswego tea, bergamot or bee-balm, is a long-time favorite of perennial gardeners. Most varieties grown are derived from *M. didyma*, common bee-balm, and *M. fistulosa*, wild bergamot, or their hybrids. Of the two, *M. fistulosa* has much less showy flowers, but has the dual advantages of better resistance to powdery mildew and greater drought tolerance. These plants thrive in moist, humus-rich, well-drained soils; consistent moisture during the summer reduces plant stress and appears to reduce powdery mildew susceptibility. Deadheading will increase flowering performance. Clumps of monarda will naturally die out at their centers, so replanting every 2-3 years helps keep plants looking their best.

The main disease affecting monarda is powdery mildew. The fungus attacks fairly early in the summer and produces white fungal growth on the upper surface of leaves of many varieties. Infected monarda leaves yellow and fall, leaving the lower stems bare of leaves. Look for varieties of monarda such as 'Jacob Cline' that have less susceptibility to powdery mildew. Fungal leaf spots are caused by *Cercospora*, *Ramularia*, and *Phyllosticta*. These generally appear as dead, brown circular areas on the leaves. Yellow leaf spotting can be caused by *Tomato spotted wilt virus* (TSWV), *Impatiens necrotic spot virus* (INSV) or other viruses. The spotting on monarda foliage from TSWV and INSV may also become necrotic. A gall forms on leaves as a result of infection by *Synchytrium*. Two species of rust (*Puccinia angustata* and *P. menthae*) also affect the leaves, producing orange-brown spores in pustules on the lower leaf surface.

In southern states, a stem rot can be caused by the fungus *Sclerotium rolfsii*, which generally forms a white mat of fungus at the base of the plant. Look for the mustard-seed-like sclerotia to identify *S. rolfsii;* before removing an infected plant, skim the surface of the soil to remove sclerotia. Infected plants in the nursery should also be discarded.

Rust on bee-balm is reddish brown and pustules form on leaf undersides.

TSWV can cause very subtle ring spots on bee-balm.

Viruses may cause yellow mottling on bee-balm foliage.

Impatiens necrotic spot virus (INSV) causes chlorotic and necrotic spotting on bee-balm leaves (left) and blackening on stems (right).

Some varieties of bee-balm are very resistant to powdery mildew while others can be heavily infected.

Myosotis

Boraginaceae

Myosotis, or forget-me-not, is a genus of short-lived perennials—but these will self-sow abundantly to renew themselves in the garden. They thrive in moist, well-drained areas, even in poor soils. Forget-me-nots are suited to partial shade, but will tolerate full sun in the North if kept moist. *M. scorpioides* can grow directly in water.

Forget-me-nots may succumb to winter moisture stresses and fail to grow out in the spring.

Symptoms of leaf blight on forget-me-not caused by a *Pseudomonas* sp.

While powdery mildew has the typical appearance of white fungal growth on the upper surface of the infected leaf, downy mildew (*Peronospora* or *Plasmopara*) forms a bluish-gray growth of spore structures on the undersides of leaves with yellow areas visible on the upper side. Forget-me-nots are also susceptible to rusts (*Puccinia*) and a white smut (*Entyloma serotinum*). Botrytis, Cercospora, or Stemphylium fungal leaf spots may occur. Thin out dense plantings after spring bloom to improve air circulation. Irrigate early in the day to allow foliage to dry before nightfall. A stem rot occurs as a result of infection by the fungus *Sclerotinia sclerotiorum*, which is favored by much the same conditions as its close relative *Botrytis cinerea*. Remove infected plants, taking particular care not to leave sclerotia (resistant fungal structures resembling mice droppings) behind in the garden. Forget-me-not is also subject to leaf blight caused by a species of *Pseudomonas* bacterium.

Narcissus

Amaryllidaceae

Narcissus, or daffodil, is a group of early blooming bulbs that are symbols of spring. A number of species and interspecific hybrids are grown. Narcissus like soil high in organic matter, and will profit from the addition of lime if soils are strongly acidic. In areas with cold winters, narcissus should be planted at least 3 inches deep. They do best in full sun or in areas with light dappled shade for part of the day. If spring is dry, irrigate so that flowers will not abort. Allow leaves to remain on the plant after flowering until they have yellowed (for about 6 weeks), to keep flowering vigororous for the next year. The length of the cold period affects the flower display: too short a cold period and the flower stems will be short and spindly. To invigorate bulbs that have just flowered poorly, fertilize with low nitrogen, high-potassium fertilizer to improve performance the next year. Divide plants when they become congested and flowering is reduced.

In most locations, narcissus is disease-free. Because it is grown in such numbers, however, in so many different areas, a number of diseases have been observed over the years. Leaf spots (and in some cases stem rots) are caused by several different fungi including species of *Botrytis, Gloeosporium, Stagonospora, Ramularia, Stromatinia,* and *Heterosporium.* The fungus *Botryotinia narcissicola* causes narcissus smolder, a disease that results in stunting and scorching of leaves emerging from diseased bulbs. Mosaics and streaking on the foliage may be caused by narcissus mosaic, narcissus yellow stripe, narcissus flower streak or any of ten other viruses known to affect this genus. Many of the narcissus viruses,

Above-ground appearance of narcissus with Stagonospora bulb rot.

Stem and bulb nematodes (*Ditylenchus dipsaci*) can infect narcissus bulbs, causing a brown ring of discoloration in cross section.

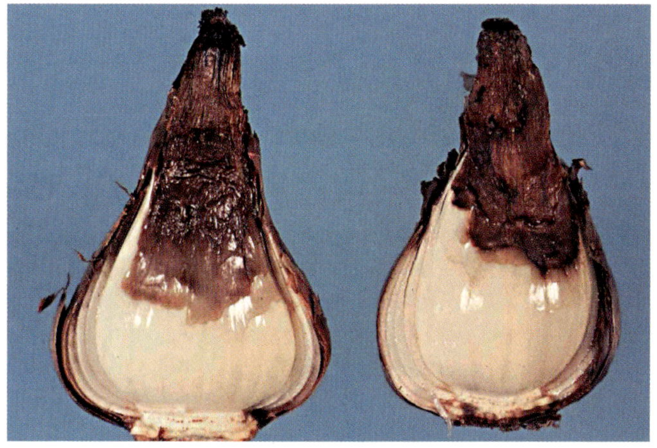

Mixed infection of narcissus bulbs with *Botrytis* and *Stagonospora* spp. can result if the bulb crop is grown under poor conditions.

including cucumber mosaic (CMV), tobacco rattle (TRV) and tobacco ringspot (TRSV), occur on other plants as well.

Bulb problems may result from the feeding of foliar (*Aphelenchoides*) and lesion (*Pratylenchus* spp.) nematodes. *Ditylenchus dipsaci*, the stem and bulb nematode, invades bulbs and produces stunted and distorted above-ground growth. Dark rings of discoloration may appear in the bulbs. The fungi *Fusarium*, *Penicillium*, *Sclerotium*, *Sclerotinia*, and *Armillaria* may also attack narcissus bulbs and cause dark or mushy, sunken areas. Do not use damaged bulbs; find a new supplier if possible. This is especially important during production of narcissus pots or in forcing.

Botryotinia narcissicola causes "smolder" on narcissus.

Sclerotia (round, black fungal survival structures) are sometimes apparent on bulbs affected by *Botrytis* sp.

Potyvirus can cause mottling on narcissus leaves.

Nepeta

Lamiaceae

Nepeta, or catmint, has many species, including the common catnip, *N. cataria*, found in herb gardens. Nepetas are full sun plants best grown in well-drained soil. In southern gardens, afternoon shade is helpful. Nepetas are grown as specimens or as edge plants. Some, such as *N. nervosa*, do well in poor soil. For a second flush of blooms, the hybrid varieties should be sheared back after flowering.

Diseases are not common on nepeta. Foliar disease possibilities include Pseudomonas and Xanthomonas leaf spots (bacterial), gray mold (caused by the fungus *Botrytis cinerea*) and fungal spots caused by species of *Ascochyta*, *Cercospora*, *Phyllosticta*, and *Septoria*. One virus that can cause mottling of the leaves or may infect plants but show no symptoms is *Cucumber mosaic virus* (CMV).

Wilt, caused by a species of the fungus *Fusarium*, is sometimes one-sided; that is, one side of the plant exhibits yellowed leaves and wilting while the remainder of the plant appears normal until later in the season, when the majority of the plant wilts and yellows. The pathogen survives in the soil even in the absence of susceptible plants. There are no fungicides that are effective in suppressing Fusarium wilt.

Oenothera

Onagraceae

Oenothera, sundrops or evening primroses, includes some North American natives such as *O. caespitosa* from the west and *O. missouriensis* (syn. *O. macrocarpa*) from the south-central United States. Evening primroses do well in sunny sites in well-drained soils; some are ideal for rock gardens or seaside gardens. *O. missouriensis* may cease blooming for a time during midsummer heat. *Oenothera* spp. may be invasive in some regions of the country but some varieties are known to be less invasive than the species. Evening primroses are attractive to Japanese beetles.

An unidentified virus in evening primrose appears as mottling and distortion of some leaves.

Evening primroses are susceptible to powdery mildew, typically seen as white fungal growth on the surface of lower leaves. Another disease of these plants is downy mildew (*Peronospora arthurii*), first noticed as angular

Spots of pigment are natural in the leaves of some evening primroses.

In these evening primrose leaves, a spherical virus was detected that was likely to have been responsible for the yellow mottling.

yellow areas on the upper leaf surface but also forming a bluish-gray growth of spore structures on the underside of the leaf. Rust (either a *Puccinia* species, which has part of its life cycle on grasses and sedges, or *Uromyces plumbarius*) appears as orange-brown clusters of spores on the undersides of the leaves. Leaf spots are caused by species of *Alternaria*, *Cercospora*, *Septoria*, *Pezizella*, and *Pestalotia*, among other fungi. A leaf gall is caused by *Synchytrium* sp., while leaf mosaic may be the result of *Tomato spotted wilt virus* (TSWV) infection.

Evening primrose is highly susceptible to infection by the bacterium *Rhodococcus fascians* during vegetative propagation. Growth abnormalities include leafy shoots and shoot proliferation.

Root rot may indicate infection by species of *Pythium*, *Phymatotrichopsis*, or *Rhizoctonia*. Tiny galls on roots are caused by the feeding of root knot nematode (*Meloidogyne*).

Brown rust pustules can develop on evening primrose on both the tops and undersides of leaves.

Opuntia

Cactaceae

Opuntia, or prickly pear, is a genus of native plants covered with tiny reddish thorns that are quite capable of penetrating gardeners' gloves. Shading, deep planting or excessive irrigation will increase the chance of disease in prickly pear, which is normally trouble-free, especially in soils with low moisture-holding capacity. The most common problems are mealybugs and scales, which are very tough to eradicate once established.

Spots on pads of prickly pear cacti may be caused by several different pests or environmental conditions. A *Phyllosticta* sp. fungus causes pad spot, which produces very black lesions. Dry rot, caused by the fungi *Phyllosticta concava* and *Mycosphaerella*, and scorch, caused by *Hendersonia opuntiae,* are also found on opuntia. In production of prickly pear and many other cacti and succulents, *Dichotophthora* and *Drechslera* spp. can cause black or brown sunken spots. The fungus *Stevensea wrightii* causes charcoal spot, a common and destructive disease on opuntia in Texas. Small spots appear first, surrounded by a ring of small raised dots that are the fruiting structures. Spots later enlarge, but remain separated. Severely infected pads or entire plants should be removed from landscapes to prevent spread of these fungi.

Cotton root rot, which occurs in several localized regions of southern Texas, New Mexico, and Arizona, is caused by the soil-dwelling fungus *Phymatotrichopsis omnivora.* Infected plants die. When plants are pulled from the soil the brown strands of the fungus can be found growing on the root surface. No control practice is available and this fungus has a wide host range.

The bacterium that causes soft rot, *Pectobacterium carotovorum,* enters tissue through natural openings and wounds. Diseased tissue is watery, black, soft, and deteriorates rapidly. The best control is to avoid wounds, treat broken surfaces right away with a copper fungicide, and avoid growing plants in places where humidity or summer rainfall is high.

Opuntia species can be attacked by root knot nematodes (*Meloidogyne*). Infected roots show small galls that can be used to identify the problem when clean, washed roots are observed. If root knot nematodes cause problems in landscape plantings, remove and destroy infected plants, and replant with non-hosts of these nematodes.

Drechslera pad spots on prickly pear are often very dark brown or black and can be covered with the black spores of the fungus.

Sunken spots caused by the fungus *Dichotophthora* sp. can be confused with insect feeding on pricky pear.

Origanum

Lamiaceae

Origanum, or oregano, although best known as a culinary herb, also includes species and varieties used as ornamentals. These plants grow best in soils with a high pH (alkaline) and good drainage. The golden varieties of *O. vulgare*, the common oregano, should be protected with some afternoon shade. *O. laevigatum* is very drought tolerant.

The leaves of oregano can be spotted by species of the fungi *Alternaria*, *Botrytis*, or *Stemphylium*, particularly under wet conditions. A rust disease may be caused by the same fungus that attacks other mints, *Puccinia menthae*. Leaf mottling and death result from infection by *Impatiens necrotic spot virus* (INSV), which is moved from plant to plant primarily by thrips. Promptly remove plants with mottled or blackened leaves from the garden or nursery. This virus attacks and damages many herbaceous perennials and annuals.

Pythium root rot causes plants to wilt and die. Avoiding overwatering and overfertilization will help to keep this problem in check.

Foliar nematodes (*Aphelenchoides*) can cause brown, vein-limited blotches on oregano leaves. Remove the blighted leaves. You can minimize splashing of soil particles onto leaves by irrigating at the soil level rather than overhead, and by mulching.

Impatiens necrotic spot virus on oregano can look like black water-soaked spots that can be confused with those caused by some fungi.

Foliar nematode on oregano can cause rather indistinct mottling and blackening on infested leaves.

Paeonia

Ranunculaceae

Paeonia, or peony, is a group of species including very showy ornamental hybrids. They prefer bright sunny locations that have well-drained, humus-rich soil and a good supply of water. Some protection from afternoon sun is appropriate in the South. The group includes common peony (*P. officinalis*) and Chinese peony (*P. lactiflora*). Fernleaf peony, *P. tenuifolia*, is more unusual, and has brittle roots that make division difficult. The woody tree peony, *P. suffruticosa,* is also commonly grown; it requires some protection from winter wind. Several cultural factors may affect the flowering of peonies. Excess nitrogen or shade, for example, may result in fewer flowers. Planting at the right depth is important for future plant performance; the buds on the rootstock should be buried only about 2 inches down for common or Chinese peonies, and the graft union of tree peonies should be set 5-6 inches below the soil. The flower buds may be blasted by thrips, fungus, or frost, as well as by by unseasonably hot temperature. In hotter climates, growing only those peonies that flower early may improve reliability of flowering. See the American Peony Society (www.americanpeonysociety.org/) for more information.

Although peonies have earned a reputation as relatively disease-free plants, several diseases can be troublesome. Botrytis blight, caused by the fungi *Botrytis cinerea* or *Botrytis paeoniae*, attacks stems, buds, and leaves, especially during prolonged periods of cloudy, rainy weather. In the spring, *Botrytis* can cause young shoots to discolor, wilt, and fall over. Later, browned buds and blighted leaves may develop masses of gray, fuzzy fungal spores that are a distinctive sign of this Botrytis blight. A disease known as leaf blotch or measles, caused by the fungus *Cladosporium paeoniae*, appears first as small, reddish spots that can coalesce to form large, irregular purple blotches on leaves and stems. Lesions are also formed on stems. Other fungi, such as species of *Gloeosporium, Alternaria, Septoria,* and *Cercospora*, can also contribute to blighting of the foliage, sometimes affecting most or all of the foliage by late summer. Powdery mildew is becoming increasingly common on peonies in the landscape; plants become coated with white spores by midsummer. A bacterial blight thought to be due to a species of *Xanthomonas* is also becoming more common: spotting may be accompanied by rings of dark red pigment or sometimes yellow haloes. Keeping these diseases in check requires growers to use adequate plant spacing and good sanitation (promptly remove and destroy spent blooms, and prune out and discard all blighted foliage in the fall), water plants at the base rather than from overhead, and remove overhanging vegetation that shades the peonies. Fungicides can be used to suppress peony foliar diseases, but are often unnecessary if proper cultural practices are followed.

Phytophthora blight is caused by a water mold, *Phytophthora cactorum*, which attacks from the soil. It can blight flowers, leaves, and stems, but unlike Botrytis blight it produces no masses of fuzzy gray spores during wet periods. Instead, infected parts become dark brown to black and somewhat leathery, and shoots may die. Crowns may also develop a dark, wet rot. Since the disease thrives in saturated soils, a preventive strategy means selecting planting sites that have good drainage. If Phytophthora blight is diagnosed, affected plants should be removed and destroyed. Peony varieties differ somewhat in susceptibility to both foliar fungal diseases and Phytophthora blight.

Foliar nematode damage on peony appears between the leaf veins as yellow, red or brown discoloration.

Botrytis blight on peony can appear as many reddish-purple spots all over the leaves.

Frost damage on peony shows as whitish patches on upper portions of the plant.

Powdery mildew can cover the entire surface of peony leaves (left) and attacks stems and petioles in severe cases (right).

White mold, caused by the fungus *Sclerotinia sclerotiorum*, can cause stem rot on peony as well as many other herbaceous plants. The entire plant may wilt, or only a portion of it. Infected areas of the stem turn a light tan color and may become withered and stringy. Under wet conditions, fluffy white fungal growth (mycelium) often appears, giving the disease its common name. If infected portions of stems are sliced open lengthwise, irregularly shaped black sclerotia will be revealed. Unfortunately, sclerotia can survive in the soil for many years, so be careful not to scatter any of them when you pull up and dispose of infected plants. Replant infested areas with species that are not susceptible to white mold, and use wide plant spacing to improve air circulation and reduce the moist conditions that favor white mold infection. A somewhat similar disease is Southern blight, caused by the soil-dwelling fungi *Sclerotium rolfsii* and *Sclerotium delphinii.* Both these fungi attack plants during warm, wet periods, causing the stems to turn water-soaked at the base, then wilt. The base of diseased stems will often show fans of thick, ropy-textured fungal mycelium and numerous, tiny, spherical sclerotia that turn from white to brick red as they mature. As for white mold, the sclerotia allow the Southern blight fungi to survive in the soil for many years. Another similarity to white mold is that these fungi can attack many different kinds of plants. Dig out and destroy diseased plants, taking care not to scatter the sclerotia, and replant only with species that are not susceptible to Southern blight.

Verticillium wilt is another occasional disease of peony whose causal fungi, *Verticillium albo-atrum* and *V. dahliae,* survive in the soil for long periods within microsclerotia. These microscopic sclerotia are not conspicuous like those of white mold or Southern blight, but they serve the same purpose of allowing the fungus to persist in contaminated soil. Symptoms caused by *Verticillium* species include wilting of shoots in the absence of damage to the crown or roots. When recently affected shoots are cut in cross section, the ring of xylem tissue in the stems often shows a brown discoloration. Remove and destroy plants with Verticillium wilt, and do not replant with peony or any other species that are susceptible to this disease, since the pathogen can survive in the soil for many years.

Virus symptoms in peony include ringspots, mosaic (a mottled pattern of light and dark green on the leaves), stunting, curling leaves, or a reduction in vigor. *Tobacco rattle virus* (TRV) causes symptoms previously referred to as peony mosaic or peony ringspot virus. *Tomato spotted wilt virus* (TSWV) produces yellow line patterns, and *Alfalfa mosaic virus* (AMV) may also cause symptoms of chlorosis. Some virus-infected peonies will grow and bloom normally except for leaf mottling or other subtle symptoms. No treatment is recommended for these mildly affected plants. If symptoms are severe, the plant should be removed and discarded.

Less commonly diagnosed diseases of peony include bacterial crown gall (pathogen: *Agrobacterium tumefaciens*), foliar nematodes, nematode injury to root systems (*Meloidogyne, Rotylenchus,* and *Ditylenchus*), fungal root rots (*Armillaria, Fusarium, Phymatotrichopsis, Rhizoctonia,* and *Thielaviopsis*), and stem wilt canker (*Coniothyrium*, occuring on tree peonies only).

Tobacco ringspot virus can infect peony alone or in combination with other viruses as shown here.

Xanthomonas blight on peony is more common in production of the crop than in the landscape.

Measles, caused by the fungus *Cladosporium paeoniae,* results in large purplish spots on infected peony leaves and stems.

Papaver

Papaveraceae

Papaver, or poppy, can contribute either vibrant hot colors or pastels to the sunny garden. Species and hybrids of Oriental poppy, *P. orientale*, are the most popular garden specimens, and these thrive in rich soil. Good soil drainage is imperative for poppies, as is a cool climate, so they are not good choices for southern U.S. gardens. They may be difficult to transplant. Poppies go dormant in mid-summer, so the foliage of adjacent plants should be used to hide them after their blooming period.

Poppies have relatively few disease problems in most of North America. Among the reported diseases are downy mildew (*Peronospora arborescens*) and the fungal diseases powdery mildew, gray mold blight (*Botrytis cinerea*), Cercospora and Septoria leaf spots, anthracnose (*Gloeosporium*), wilt (*Verticillium albo-atrum*), and root and stem rot (*Rhizoctonia*). Additional poppy diseases include bacterial blight (*Xanthomonas*) and diseases caused by nematodes (foliar nematode, *Aphelenchoides* species, and root knot nematode, *Meloidogyne*). To avoid encouraging root and foliar diseases, be careful not to overwater poppies. In nursery cultivation in containers, poppies require a growing medium with excellent drainage.

Xanthomonas leaf spot shows angular dark brown spots on poppy leaves.

The result of root rot on poppy is often stunting, yellowing and overall decline (plant on right).

Bacterial ooze on a poppy flower bud blackened by a *Xanthomonas* sp.

Patrinia

Valerianaceae

Patrinia species grow in sunny or shaded, well-drained sites, in moist soil high in organic matter. In the southern United States they benefit from some afternoon shade. The drought-tolerant scabious patrinia, *P. scabiosaefolia*, also known as golden valerian, is tolerant of heat and humidity and thus does well in the South.

Patrinia species are relatively recent introductions to North America, and there are few reports of disease problems on them. In Asia, the native range of these perennials, powdery mildew and rusts are the most commonly reported diseases. *Patrinia* species are alternate hosts of *Puccinia hemerocallidis*, the fungus that causes daylily rust. Daylily rust is a destructive disease of *Hemerocallis* that appeared in the United States for the first time in 2000. However, no rust-infected *Patrinia* plants have been detected in the United States as of 2008. The spread of rust from daylily to daylily is not dependent on *Patrinia*, but the widespread presence of *Patrinia* could potentially allow the fungus to complete its life cycle, thereby increasing the risk of development of new, more aggressive strains, with enhanced ability to develop fungicide resistance.

Daylily rust symptoms on patrinia.

Penstemon

Scrophulariaceae

Penstemon, or beardtongue, contains primarily North American natives from both East and West. Garden plants are often European hybrids developed from eastern North American species. Most penstemons need dry, sunny, well-drained sites, but some of the eastern species are less demanding of dry conditions. Mulching with gravel rather than organic matter is recommended to prevent crown rot. Although generally grown in full sun, light shade may help in hot climates. *P. barbatus*, common beaded tongue, and *P. digitalis*, smooth white penstemon, are heat tolerant and winter hardy.

Leaf spots (caused by species of *Cercospora*, *Septoria*, *Cercosporella, Phyllosticta, Ramularia,* and other fungi) cause sharply delimited brown spots on leaves. Leaf spots usually are favored by wet conditions and may become important if a large number of spots appear or if they start to coalesce. Cultural management calls for watering plants at the base rather than overhead. If overhead watering is the only option, it should be done in late morning to midday to speed up drying of foliage. Control can also be achieved with the use of labeled fungicides that are applied as soon as early symptoms are visible.

Powdery mildews produce patches of white fungal mycelium and spores on leaf surfaces. Severely infected leaves can also be twisted and yellowed. Powdery mildew is favored by humid periods without rainfall. Small black overwintering structures called chasmothecia are often found on the affected leaves. Control may also be achieved with use of labeled fungicides, starting when symptoms are first visible.

Rust (*Puccinia andropogonis*) symptoms include orange spores in powdery pustules on leaves (mainly on the

A *Phyllosticta* species can cause wet, black spots on penstemon.

Pythium root rot on penstemon liners can appear as yellowing and marginal browning of foliage.

undersides) or stems. Spots may appear slightly sunken when viewed from the top of leaves. Surrounding tissue is discolored and yellowed, and in severe infection plants are stunted. Control of rust may also be achieved with the use of labeled fungicides applied as soon as symptoms are visible. Be sure to add a wetting agent to the fungicide for optimal benefit.

Stem rot, caused by the soil-dwelling fungus *Sclerotium rolfsii*, causes yellowing of lower leaves, followed by wilting and death of the rest of the plant. A white cottony mass of mycelium growing around the crown or on the soil near the crown distinguishes this crown rot from others. Within this fungus growth may be found spherical, pinhead-size, white to buff to reddish brown, seed-like sclerotia. Management options include removing and destroying all infected plant parts, and/or removing the topsoil around the plant. New soil should be used to replace sclerotium-infested soil.

Root rot, caused by species of *Fusarium*, *Pythium* and *Rhizoctonia* fungi, has been reported from poorly drained soils. Selection of well-drained soils and potting media should minimize the risk of root rot diseases. The stem and bulb nematodes (*Ditylenchus*) attack almost all parts of penstemon, causing leaf lesions, yellowing, necrosis and leaf drop, bud malformation, plant stunting, stem swelling or galls, and often leading to secondary fungal infections of necrotic tissues. The nematodes live and move within plant tissues, or in water films on plant surfaces. Reducing leaf moisture and removing infected tissues, debris or plants is important. Root knot nematodes (*Meloidogyne* spp.) cause small swellings and other deformations of the root system, and reduce plant vigor. Remove and destroy affected plants, and replace with non-hosts of root-knot nematode.

Stunted, strapped leaves are symptoms of *Cucumber mosaic virus* (CMV) on penstemon.

Impatiens necrotic spot virus (INSV) on penstemon appears as purple and red mottling on infected leaves.

Magenta spots with yellow haloes may form on the upper surface of penstemon leaves infected by a rust fungus.

Penstemon rust on the underside of an infected leaf.

Pentas

Rubiaceae

Pentas or star flower are subshrubs that are not frost tolerant, but they may be grown as perennials in the Southwest, Southeast, and on the west coast of the United States. They do well in sunny sites in well-drained fertile soil. Pentas are highly attractive to butterflies and hummingbirds.

Star flowers have few serious disease problems. The gray mold fungus (*Botrytis cinerea*) can attack leaves and stems under prolonged high-humidity conditions. Pythium root rot (*Pythium*) appears as soft, brown, mushy discoloration of roots. To reduce the risk of this disease, select well-drained sites or modify the site to enhance drainage. Tan, brown, or black diseased areas at the soil line are indications of Rhizoctonia root and stem rot (*Rhizoctonia solani*). To avoid problems with Rhizoctonia diseases, keep mulch away from the stem base and never propagate from infected plants or those that look unhealthy. Other fungal diseases reported on pentas include Southern blight (*Sclerotium rolfsii*), powdery mildew, and leaf spots (*Alternaria, Cercospora,* and *Phyllosticta* species).

Alternaria leaf spot on star flower cuttings.

Pseudomonas leaf spot can start on star flower cuttings during the rooting process.

Rhizoctonia cutting rot and web blight is common on star flower in production and in the landscape in the southern United States.

Mottling on star flower may be caused by infection by a virus, such as *Cucumber mosaic virus* (CMV).

Perovskia

Lamiaceae

One species of *Perovskia* is used extensively in American gardens: *P. atriplicifolia*, Russian sage. Russian sage is a subshrub with aromatic foliage. It grows well in full sun, in well-drained soil (good drainage is helpful for avoiding injury during wet winters). Plants will naturally fall over in the garden, so staking is sometimes desirable. These plants grow rapidly and can become very large. They do not perform as well in hot, humid southern climates as they do in more northern areas. Perovskias are drought and salt tolerant.

Russian sage is often found on lists of plants that have very few disease problems. In moist climates and years, however, a fungal leaf spot (*Cylindrosporium*) and a stem canker disease (*Phoma*) have been reported. Cultural practices that will help to reduce the risk of fungal foliar diseases include spacing plants to insure good air movement and watering at the base rather than overhead; these practices will minimize the time that foliage is wet.

Phlox

Polemoniaceae

Phlox is a genus of North American natives, some of which are annuals. The best known perennial types are creeping moss phlox, *P. subulata,* and garden phlox, *P. paniculata*. They are grown in well-drained soils. Full sun is best for the upright forms, and *P. subulata* can also be grown in sun, but both *P. divaricata*, the woodland phlox, and *P. subulata* also grow well in light shade. The foliage of phlox will deteriorate in droughty conditions, and spider mites will add to the problem for *P. paniculata*. Deadhead *P. paniculata* and *P. maculata* to encourage more flowers. *P. maculata*, spotted phlox, is less prone to powdery mildew.

Powdery mildew is overwhelmingly the most common disease of *Phlox paniculata*. It appears as conspicuous powdery white spots on foliage and stems. In severe cases, older leaves may be entirely coated with powdery mildew, and flowers may be affected as well. Maintaining good air circulation and minimizing shading can help to suppress powdery mildew. Watering at the base of plants rather than using overhead sprinklers helps to keep the plant canopy dry. Another strategy is to move the phlox to a sunnier location in the landscape. A key preventive practice is to select varieties that are resistant to the disease; many garden phlox varieties have moderate or high levels of resistance. Several widely available fungicides are labeled for powdery mildew control on phlox, but sprays must begin soon after the first spots are seen in order to be effective.

Fungal leaf spot diseases, caused by *Ascochyta, Cercospora, Macrophoma, Phyllosticta, Septoria, Volutella, Ramularia* and *Stemphylium*, can appear on lower leaves, especially after extended wet periods, but are rarely damaging enough to warrant treatment. Keeping foliage dry and providing adequate spacing will help to suppress fungal leaf spot diseases.

Several other less common diseases have been reported from phlox; these may be damaging in some locations and years. Among the diseases caused by fungi are gray mold blight (caused by *Botrytis cinerea*), Southern blight crown rot (*Sclerotium rolfsii*), stem blight (*Pyrenochaeta*), downy mildew (*Peronospora*), stem canker (*Colletotrichum*), charcoal rot (*Macrophomina*), root rots (*Thielaviopsis basicola, Phymatotrichopsis*), rusts (*Puccinia* and *Uromyces*), and wilt (*Verticillium albo-atrum* and *Fusarium*).

Bacterial diseases include crown gall (*Agrobacterium tumefaciens*), and phytoplasma diseases include aster yellows and Western aster yellows. Bacterial leaf spot on phlox may be caused by *Pseudomonas cichorii*. Nematodes that attack phlox include foliar nematodes (*Aphelenchoides* spp.), stem and bulb nematode (*Ditylenchus* sp.), and root knot nematodes (*Meloidogyne* spp.). *Ditylenchus* are particularly destructive to phlox and can eliminate a planting within one year. Virus diseases reported include: from *Phlox stolonifera, Alternanthera mosaic virus* (AltMV), *Phlox virus S* and *Tobacco ringspot virus* (TRSV); from *Phlox drummondi, Tobacco mosaic virus* (TMV); and from *Phlox divaricata* and *P. paniculata, Tobacco rattle virus* (TRV).

Chlorotic ring spots surrounded by black lines appear on phlox infected with a potyvirus.

Foliar nematode can result in severe stunting on infected phlox (plant on the right).

Downy mildew on phlox can appear as if plants have been sprinkled with salt or sugar.

Fusarium stem rot on phlox may show as one-sided death with yellowing and browning.

Stem and bulb nematode on phlox infests terminals and causes yellowing and browning. Note characteristic lesions at the base of the leaves.

Chlorotic mottling appears on phlox infected with *Tobacco rattle virus* (TRV).

Powdery mildew appears on all surfaces of phlox. At the end of the growing season, the fungus develops black specks (chasmothecia) that are its survival stage.

Phlox infected with the devastating black root rot disease show truly black roots.

Septoria leaf spots on phlox are round and bordered by a reddish edge.

Phormium

Agavaceae

Phormium is called New Zealand flax or flax lily and has many varieties that are popular specimen plants in gardens. Phormium should be grown in sunny locations in moist, fertile, well-drained soil. Deep, dry mulches will help with winter survival where hardiness is marginal.

Several fungal leaf spot diseases have been reported, caused by *Phyllosticta*, *Mycosphaerella*, *Cercospora phormii*, *Cercosporidium*, *Glomerella cingulata*, *Gloeosporium phomiforme*, and *Macrophoma*. Recommended practices to reduce risk of leaf spot diseases emphasize adequate plant spacing, and watering at the base rather than overhead, to minimize the length of time that foliage remains wet.

Flax lily occasionally experiences problems with crown rot, caused by *Pythium* sp., *Rhizoctonia*, *Fusarium oxysporum*, and *Armillaria*. Cultural practices to minimize the risk of crown and root rots include selecting well-drained soils, avoiding moisture extremes (excessively dry or wet), and using labeled fungicides; however, fungicides are unlikely to be consistently effective unless the disease-suppressive cultural practices are also followed.

Cercosporidium leaf spot on New Zealand flax can occur in both production and the landscape and looks like tiny black speckles.

Fusarium cutting rot is common and devastating in propagation of many New Zealand flax varieties.

Phyllosticta leaf spots are elliptical and bordered by red on New Zealand flax.

Black chevron patterns caused by an unidentified virus on New Zealand flax.

Physalis

Solanaceae

Physalis, or Chinese lantern, is a genus of solanaceous plants that includes the tomatillo. The common Chinese lantern, *P. alkekengi*, is the only ornamental of note. It spreads aggressively. Chinese lanterns are grown in full sun or partial shade, in well-drained soil. If plants are protected from drought stress and not overfertilized, they will form large, attractive fruits.

Chinese lantern can be attacked by several fungal leaf spots caused by *Alternaria*, *Phyllosticta* species, and *Phyllosticta physaleos*. White smut, caused by the fungus *Entyloma polysporum*, produces white to yellowish green spots, up to ¼ inch in diameter, that develop light to dark brown centers. These spots may later drop out, leaving ragged holes in the leaves. Thorough sanitation practices can manage leaf spot and white smut. Infected leaves or severely infected plants should be removed and destroyed. Avoid crowding plants, and water plants at the base if possible, rather than from overhead, to reduce the time that leaves remain wet.

Verticillium wilt caused by *Verticillium albo-atrum* may occur on Chinese lantern. The yellowing, withering, and plant death observed with this disease starts as a root infection. As a general preventive practice, it is helpful to avoid planting Chinese lantern where any other solanaceous ornamental or vegetable (for example, tomato, pepper, eggplant) has been grown previously, since fungal pathogens that affect this family can build up in the soil on these plants. Similarly, if physalis has been attacked by either *Verticillium* or *Fusarium*, avoid replanting this area with any solanaceous plant.

Chinese lantern can also be attacked by several viruses, including cucumber mosaic (CMV), tobacco mosaic (TMV), physalis mottle (PhyMV), and henbane mosaic (HMV.) Virus infections result in a mottled pattern of light green and normal green color on leaves, and leaf malformation in severe cases. Virus-infected plants should be removed and destroyed.

Physostegia

Lamiaceae

Physostegia, or obedient plant, has one species, *P. virginiana*, whose varieties are grown as ornamentals in sunny or partly shaded sites. *P. virginiana* is native to the eastern United States; despite its common name, it is an aggressively spreading plant that requires attention from the gardener to be kept in its place. Plants should be grown in moist, well-drained soils, preferably with an acid pH. Obedient plants will grow best in fertile soil; they may show signs of nitrogen deficiency (pale green leaves) unless given supplemental fertilizer.

Rust, caused by *Puccinia physostegiae*, is the most commonly observed disease on obedient plant. Additional diseases that occasionally appear on foliage include downy mildew caused by *Plasmopara cephalophora* and leaf spots caused by *Mycosphaerella physostegiae* and *Septoria physostegiae*. To reduce the risk of these diseases, it is helpful to minimize the time that foliage is wet by spacing plants to encourage air movement, and by watering at the soil surface rather than from overhead.

Southern blight and crown rot, caused by the fungus *Sclerotium rolfsii*, and stem rot, caused by *Sclerotinia sclerotiorum*, occasionally attack obedient plants, causing them to wilt and die. White, fluffy strands of fungus are visible at the base of diseased plants. *S. sclerotiorum* produces hard, black, lumpy sclerotia, up to ½ inch long, within or outside stems; the sclerotia produced inside stems can be observed when stems are cut lengthwise. *S. rolfsii* produces small, round, pinhead-size sclerotia that darken from cream-colored to tan and reddish brown as they develop. Diseased plants should be removed and destroyed, taking care not to scatter sclerotia. Soil in the immediate area of the diseased plants can be dug out to a depth of several inches and replaced with uncontaminated soil. Replant species that are not known to be hosts of either fungus. Several fungicides are labeled for control of these diseases.

Some virus diseases have been reported on obedient plant, including *Tomato spotted wilt virus* (TSWV), *Alfalfa mosaic virus* (AMV) and an unidentified member of the potexvirus group. Virus symptoms may appear as mottling of the foliage, sometimes accompanied by distortion of leaf shape. Virus-infected plants should be pulled up and destroyed.

Alfalfa mosaic virus (AMV) on obedient plant causes a very general yellow blotchiness on infected leaves.

Platycodon

Campanulaceae

Platycodon, or balloon flower, includes only one species that is used as an ornamental: *P. grandiflorus*. This is a long-lived perennial, to be grown in full sun; only in southern U.S. gardens should it have some afternoon shade. Balloon flowers do best in moist, well drained, fertile soils that are high in organic matter. This plant is late to emerge in the spring. It is not necessary to divide clumps of platycodon, and the plant does not recover well after division.

Balloon flowers are bothered by few diseases. The main threats are from soilborne pathogens, especially blight caused by *Phytophthora cactorum*, root rot caused by *Rhizoctonia solani* and—in localized areas of the extreme southern United States—*Phymatotrichopsis omnivora*. Avoiding heavy soils and poorly drained sites, as well as care in preventing overwatering, should minimize the risk of these diseases. Overhead irrigation can also lead to Alternaria leaf spot, which causes black, rapidly growing spots during nursery production. During greenhouse production, keep thrips populations under control—these vector the *Impatiens necrotic spot virus* (INSV), which can infect balloon flower.

Alternaria causes black, rapidly growing spots on balloon flower during production.

Sclerotinia blight sometimes shows the lumpy sclerotia of the fungus on stems of infected balloon flower.

Greenhouse-grown balloon flower shows mottling caused by *Impatiens necrotic spot virus* (INSV).

Podophyllum

Berberidaceae

Podophyllum, or mayapple, is a perfect plant for shady woodland gardens. The species used in the ornamental trade include the native *P. peltatum* and the Asiatic *P. hexandrum*. Of the two, *P. peltatum* can tolerate drier soils. Both plants prefer moist soil, high in organic matter, and light to full shade. A mulch can be used for winter protection.

Although mayapples are usually healthy, one widely occurring disease is a rust caused by the fungus *Puccinia podophylli*. The undersides of affected leaves are covered with pustules. A widespread leaf blight disease is caused by the fungus *Septotinia podophyllina*. Fungal leaf spots can be caused by *Cercospora podophylli*, *Glomerella cingulata*, *Pezizella oenotherae*, *Phyllosticta podophylli*, and *Vermicularia podophylli*. Gray mold blight, caused by the fungus *Botrytis cinerea*, can also attack the leaves and flowers. The best way to protect mayapples against these foliar diseases is to adjust watering practices to minimize the time that the leaves are wet.

A stem rot can be caused by the soilborne fungus *Rhizoctonia* sp. Avoiding soil moisture extremes may help to reduce the risk of this disease. Irrigate thoroughly, but only as needed.

Browning and collapse of mayapple leaves may signal infection by *Sclerotium rolfsii*, which starts at the stem base.

Mayapple rust causes pale patches on leaves, and shows orange pustules on leaf undersides.

Polemonium

Polemoniaceae

Polemonium, or Jacob's Ladder, includes both native and non-native species that are best suited for cooler climates. The European native, *P. caeruleum*, is available in a number of varieties. The creeping polemonium, *P. reptans*, with varieties 'Alba' and 'Blue Pearl', does well in the eastern United States; a number of other species are more long-lasting in western gardens. Polemoniums are grown either in full sun or partial shade, depending on the species, in moist, well-drained areas.

Rust diseases, caused by the fungi *Puccinia gulosa*, *P. polemonii*, and *Uromyces acuminatus* var. *polemonii*, have been widely reported on *Polemonium*, on which they cause pustules of reddish brown spores on foliage. Powdery mildew causes granular-appearing white patches on foliage, and can distort leaf shape. Leaf spot symptoms can result from attack by species of *Cercospora* and *Septoria* fungi. Helpful cultural practices to minimize these diseases include insuring adequate spacing for rapid drying of the foliage, watering at the soil line to avoid wetting leaves, and watering early in the day if overhead watering is unavoidable.

Wilting can be caused the soilborne fungi *Fusarium* and *Verticillium albo-atrum*. The water molds *Pythium* and *Phytophthora* may also attack polemonium root systems. Selecting well-drained planting sites and avoiding moisture extremes can help to lower the risk of wilt and root rot diseases. If fungal wilts are diagnosed, remove and destroy the affected plants, and replant with species that are not susceptible to these diseases.

Tomato spotted wilt (TSWV) is the only virus disease reported for polemonium. Infected plants may display mottled yellow-green discoloration of leaves or light-colored rings on foliage. Infected plants should be removed and destroyed to prevent spread by the insects that are vectors of TSWV, the western flower thrips.

Phytophthora crown rot caused the death of this polemonium.

Collapse of polemonium plants due to Pythium and Fusarium root rot.

Powdery mildew on polemonium is usually found as white patches on the upper leaf surface and on stems.

Polemonium powdery mildew can appear as purple or red discoloration at times.

Polygonatum

Liliaceae

Polygonatum, or Solomon's seal, is grown for the beauty of the plant's form more than for its spring flowers. Both dwarf and large species are available for the shady, moist garden. Partial shade is sufficient in most areas, but the plants can tolerate full shade. Moist, well-drained soil high in organic matter is desirable. Two species native to the United States are often grown: *P. biflorum*, small Solomon's seal and *P. commutatum*, great Solomon's seal. Other common garden species are from Asia.

Polygonatums are generally disease-free. Two rusts, caused by the fungi *Puccinia sessilis* and *Uromyces acuminatus* var. *marginatus*, are the most widely reported diseases. Both are alternate-host rusts, which means that they must have another particular type of plant (the alternate host) nearby in order to complete their life cycles. The alternate host of *P. sessilis* is *Phalaris* sp. (ribbon grass and reed canarygrass), whereas the alternate host of *U. acuminatus* var. *marginatus* is the marsh grass *Spartina*. One way to lessen damage caused by alternate-host rusts is to eliminate one of the two hosts from the vicinity of the more desirable host plant. The farther away the hosts are from each other, the less likely it is that they will trade rust spores back and forth. For rust as for fungal leaf spots of Solomon's seal (caused by *Colletotrichum liliacearum* and *Sphaeropsis cruenta*), reducing the time that foliage is wet can minimize the risk of disease. Fungicides can also suppress both types of foliar diseases, but are seldom needed. Additional diseases reported for Solomon's seal include leaf smut (caused by the fungus *Urocystis colchici*) and an unidentified virus that causes mosaic symptoms.

Rust on Solomon's seal forms pustules on leaf undersides (left) and yellow leaf spots (right).

Primula

Primulaceae

Primula, or primrose, has many species, and is an important garden plant. Cultural requirements vary within the group. Most primroses do best with cool summers, partly shaded settings, and well-drained soil that is consistently moist. Some need higher pH (alkaline) soil; others require acid soil. Compared to some *Primula* species, the hybrid *P. polyantha* has less need for moisture, but it is not drought tolerant. The English primrose, *P. vulgaris*, has better heat tolerance than many of the others, and also does not require constantly moist soil. Auricula primrose, *P. auricula*, may have a natural coating of mealy-looking leaf wax which closely resembles powdery mildew, so don't be fooled. This can appear on leaves, flower stems, and calyxes. In botanical language, primroses with this trait are said to be "farinaceous".

Gray mold, caused by the fungus *Botrytis cinerea,* is the most common disease problem of primroses in greenhouses, and occasionally appears in outdoor situations. Leaves, stems, and flower buds are blighted. The fungus forms masses of grayish brown, fuzzy fungal growth on these blighted areas under very humid conditions. Reducing humidity and increasing air flow around plants will suppress activity of gray mold. Powdery mildew is favored by some of the same environmental conditions that *B. cinerea* prefers, so reducing humidity and increasing air flow are also important controls for this disease.

Among the fungi that cause leaf spot diseases on primrose are *Ascochyta primulae, Asteroma garretrianum* (black spot), *Cercospora primulae, Colletotrichum primulae, Mycosphaerella*, and *Ramularia primulae.* A bacterial leaf spot caused by *Pseudomonas primulae* has also been reported. Cultural practices to suppress leaf spot include spacing plants widely enough apart to allow rapid drying, and watering at the base rather than overhead. If overhead irrigation is used, watering in the morning is preferable, since it allows the foliage to dry more rapidly than watering in late afternoon or evening.

Root rot diseases of primrose are caused by *Pythium irregulare, Rhizoctonia solani,* and (in the extreme southern U.S. only) *Phymatotrichopsis omnivora.* In general, the risk of root rots can be lessened by avoiding extremes of soil moisture (too wet or too dry). Drench treatments of fungicides may be helpful in some situations, but proper cultural management is essential even when chemicals are used.

Nematode problems that occur occasionally on primose include both leaf and stem nematode (*Ditylenchus dipsaci*) and root knot nematodes (*Meloidogyne* spp.). To minimize the risk of leaf and stem nematodes, it is helpful to mulch and water in a manner that avoids splashing water onto foliage. If root knot nematodes are causing damage, it is advisable to replace injured plants with species that are not hosts of these nematodes.

Viruses that attack primroses include primrose mosaic (PrMV), primula mottle (PrMoV), cucumber mosaic (CMV), tobacco necrosis (TNV), tomato spotted wilt (TSWV), and impatiens necrotic spot (INSV). Virus symptoms can include mottling of leaves and distortion of leaves and stems. Virus control efforts in greenhouses focus on suppressing the insects that spread most of these viruses. In the landscape, virus-infected plants should be removed so that they do not become sources from which insects can spread the problem to other plants. Aster yellows, caused by a phytoplasma, is spread by certain species of leafhoppers. Symptoms include stunting, excessive branching, and yellow, strap-shaped leaves, as well as flower abnormalities. Management for aster yellows focuses on suppressing leafhoppers and keeping weeds under control.

Pseudomonas leaf spot on primrose.

The dusty gray spores of *Botrytis cinerea* form readily on blighted primrose.

Upper and lower leaf surfaces of primula infected by a downy mildew. Note the downy sporulation on the lower surface.

Pulmonaria

Boraginaceae

Pulmonaria, the lungworts, are named after a fancied resemblance to ulcerated human lungs, which once led to their being prescribed as a remedy for lung ailments. Many pulmonarias come complete with spots that are natural to them. Pulmonarias are good for areas in partial shade, spreading to form a ground cover in moist soils with good drainage. *P. officinalis*, Jerusalem cowslip, may be grown in full sun. Poor drainage in winter is to be avoided to prevent plant loss. Foliage degenerates if plants are exposed to drought or over-irrigation. The leaves can be cut back to renew the foliage in midsummer.

Pulmonarias are susceptible to powdery mildew, which results in patches of white fungal growth on leaves in late summer. Siting plants where they will get morning sun can help to suppress powdery mildew. Because several varieties are resistant to the disease, you may scout garden centers in late summer to observe which pulmonarias are powdery mildew resistant in your area. Other foliage symptoms may be caused by virus infection: deformed leaves, mosaic and leaf desiccation have been associated with *Alfalfa mosaic virus* (AMV) in pulmonaria; *Cucumber mosaic virus* (CMV) and an unidentified filamentous virus have also been found in this host.

A crown rot caused by the soilborne fungus *Sclerotium delphinii* can kill pulmonaria during warm, wet periods. Look for masses of white fungal growth and spherical, pinhead-size, white to reddish brown sclerotia (survival structures) at the base of collapsed plants. This fungus lacks spores; instead, it grows across the soil surface from plant to plant, attacking at the soil line. Affected plants should be removed and discarded, including as many of the sclerotia as possible. Soil in the vicinity of the sick plants can be removed and replaced. Replanting into infested areas should be with plants that are not known to be hosts of this fungus or its close relative, *Sclerotium rolfsii*.

Virus-like symptoms on lungwort include mottling and ring spots.

Mixed infection can be common on foliage in wet conditions—shown here, Alternaria and Colletotrichum leaf spots on lungwort.

Crown rot on lungwort can be caused by *Sclerotium rolfsii.*

Powdery mildew has coated the older leaves of this lungwort plant.

Ranunculus

Ranunculaceae

Ranunculus, or buttercup, is a large genus that encompasses many species, including members with several different kinds of root system. The familiar potted plant, *Ranunculus asiaticus*, Persian buttercup, is not hardy in most of the United States; its tuberous roots must be dug and stored for winter unless in zones 8-10. Persian buttercup flowers best in areas with cool springs, and grows in well-drained soils as long as it is provided with some shade. Two stoloniferous buttercups–lesser celandine, *R. ficaria*, and the creeping buttercup, *R. repens*–are both extremely aggressive ground covers. *R. montanus*, mountain buttercup, is a less-aggressive ground cover. The widely naturalized *R. acris*, meadow buttercup, has been developed into slower-spreading varieties with doubled flowers. Full sun and moist conditions result in the most luxuriant growth of ranunculus; partial shade helps to keep these plants in check. Foliage naturally deteriorates on *R. ficaria* when it goes dormant in summer.

Powdery mildew is common on ranunculus. Symptoms include white, granular-appearing patches on foliage, and leaf distortion. Several rust fungi occasionally attack ranunculus, including *Puccinia andina, P. eatoniae* var. *ranunculi, P. recondita, Uromyces dactylidis*, and *U. jonesii.* To reduce rust problems, an option for the alternate-host rusts (*P. eatoniae* var. *ranunculi* and *P. recondita*) is to establish the buttercup planting as far as possible from the alternate host plants (wedgegrass and barley, respectively). To fight powdery mildew and rust, it is helpful to increase airflow among plants and to water early in the day to allow plants to dry quickly. Several fungicides are labeled for powdery mildew and rust control, but will not replace the need for proper cultural practices.

Ranunculus occasionally develops downy mildew, caused by the fungus-like organism *Peronospora ficariae.* The most distinctive symptoms of downy mildew are tufts of white to light gray fuzz (growth of the downy mildew organism) on the undersides of leaves. Accompanying symptoms may include yellowish or pale green foliage, downward curling of leaves, leaf distortion, leaves that are small and/or discolored as they emerge, failure to form flower buds, and stunted plants.

The humid, wet environments that stimulate downy mildew also favor attack by the gray mold fungus, *Botrytis cinerea.* Gray mold results in blighted leaves or stems, and produces masses of fluffy, ash-gray fungal growth on diseased areas during prolonged wet periods. The same cultural practices described for powdery mildew will help to reduce risk of gray mold. Fungicides that are effective and labeled against gray mold often differ from those effective against downy mildew, however.

Pythium root rot and damping-off can occur in some instances on ranunculus.

Fungal leaf spots that can show up on ranunculus are caused by species of *Ascochyta, Cercospora, Cylindrosporium, Didymaria, Fabraea, Ovularia, Ramularia, Septocylindrium,* and *Septoria*. The same cultural practices described above, to limit the duration of periods when foliage is wet, can help to suppress these leaf spots. Some fungi attack the stem base rather than the leaves; *Sclerotinia sclerotiorum*, for example, causes white mold, a disease that is usually fatal to infected plants. White mold is fairly easily recognized by the white fungal growth on the lower stem, in which black sclerotia roughly the size and shape of mouse droppings may be imbedded.

Bacterial leaf spot, caused by *Xanthomonas campestris,* is common in field-grown and containerized production in some parts of the country. The bacterium has also been found in seeds and tubers, suggesting that it may spread in contaminated planting materials. Aster yellows, caused by a specialized type of bacterium called a phytoplasma, results in severely deformed plant growth. It is carried from plant to plant by leafhoppers, particularly the aster leafhopper. Yellows-infected plants should be removed and destroyed; insecticides can be used preventively to deter the leafhoppers.

Virus diseases reported from some species of *Ranunculus* include impatiens necrotic spot (INSV), tomato spotted wilt (TSWV), beet curly top (BCTV), and ranunculus mottle (RanMoV), plus ranunculus mosaic and other unclassified potyviruses. Virus symptoms can vary greatly depending on the virus, plant age, environmental conditions, and other factors, but may include wilting, stem death, stunting, yellowing, poor flowering, sunken spots on leaves, and ringspots. Since many viruses are carried by specific types of insects—for example, INSV and TSWV are carried by thrips—virus control often focuses on monitoring, excluding, and/or killing these insects.

Xanthomonas leaf spot on ranunuclus is sometimes angular and sometimes round in shape.

Powdery mildew shows typical white patches on ranunculus.

Impatiens necrotic spot virus (INSV) is often found on ranunculus with brown or black markings on leaves and petioles.

Rosmarinus

Lamiaceae

Rosmarinus, or rosemary, includes two species of shrubs. The common rosemary, *R. officinalis*, is well known as a culinary herb, but this plant may also delight the eye, so it is often used in perennial plantings. Rosemary is hardy to Zone 7, or Zone 6 with protection; new growth will come back from the surviving root systems in cold areas within its range. Wet soils in winter can cause plant loss, so good drainage is essential. Grow rosemary in full sun, in either acidic or basic soils of low to moderate fertility.

Rosemary with mushy brown roots may have Pythium or Phytophthora root rot.

Although rosemary is often disease-free in the landscape, powdery mildew can be severe on plants that have been moved inside to overwinter, and also during springtime greenhouse production. Powdery mildew will appear on foliage as patches of white, grainy-looking fungus. To reduce greenhouse problems with powdery mildew, keep relative humidity low, circulate the air and ventilate in the evening to prevent condensation on the plants. Botrytis blight (*Botrytis cinerea*) on both leaves and stems can also occur in greenhouses under very moist conditions with poor air circulation; again, keeping the relative humidity low will help greatly to suppress the disease. Fungicide sprays registered for use on herbs are also an option against both powdery mildew and Botrytis blight, but keeping moisture under control is an essential part of integrated management of these diseases.

Downy mildew symptoms can include leaf yellowing or mottling; blighted leaves turn purplish, then brown and die. Downy mildews usually develop during cool, wet conditions with high relative humidity. Prolonged periods of leaf wetness promote the disease. Spores are easily

Powdery mildew on rosemary happens mainly in the late winter and early spring.

Rosemary may show a Xanthomonas leaf spot.

spread by wind and splashing water. In the greenhouse, keep relative humidity below 85%.

Bacterial blights of rosemary can be caused by a *Xanthomonas* or *Pseudomonas syringae.* These diseases, though not common, can spread rapidly outdoors in rainy weather, discoloring and killing the foliage. Good general practices to manage bacterial diseases are aimed at prevention; they include removing and destroying infected plants, replanting with non-host plants, and avoiding overhead irrigation.

Other rosemary diseases include fungal root rots caused by *Phymatotrichopsis omnivorum*, *Thielaviopsis basicola*, *Phytophthora*, and *Pythium*, and aerial blight and root and stem rot caused by the fungus *Rhizoctonia solani.* White mold, caused by the fungus *Sclerotinia sclerotiorum,* produces conspicuous masses of white mycelium at the stem base that help to identify this problem. The sclerotia of the fungus may also be present, appearing as black lumps buried in the white mold.

Web blight caused by *Rhizoctonia solani* is common during the summer in southern U.S. gardens and in greenhouse production.

Rudbeckia

Asteraceae

Rudbeckia, the coneflowers, are North American natives. The various species grow in soil of normal fertility in full sun to part shade. Some are drought tolerant (such as *R. hirta*), and others prefer moist soils, such as *R. laciniata.* They are long-flowering, strong performers in the perennial garden.

Several fungi cause leaf spot diseases on rudbeckia, including species of *Septoria, Ramularia, Cercospora, Cercospora, Corynespora, Phyllosticta, Alternaria, Cylindrocladium* and *Colletotrichum*. The most common of these is *Septoria rudbeckiae,* which causes small, dark brown spots that enlarge to ⅛ to ¼ inch in diameter. These spots start on lower leaves and gradually spread upward. With overhead irrigation, Septoria leaf spots may coalesce to cover the leaf surface. To manage the fungal leaf spot diseases, remove infected leaves to reduce inoculum levels and space plants to improve air flow around them. Avoid overhead watering or water in late morning or at midday so that foliage may dry quickly. Applications of fungicides may protect new growth and reduce the spread of Septoria leaf spot. Preventive treatments are suggested if the disease has been a problem in previous seasons. Both *Pseudomonas* (especially *P. cichorii*) and *Xanthomonas* bacteria produce angular, brown spots on the leaves that may engulf the entire leaf, appearing first on the lower leaves. Angular leaf spot due to *Xanthomonas* appears most often on 'Goldsturm' and can be avoided by switching to another variety. In the fall after the first hard freeze, remove all above-ground plant tissue. In the spring, sprays with a copper-based bactericide can minimize the severity of infection. It is important to remember to avoid overhead watering since these bacteria are easily spread in splashing water.

The presence of patches of powdery white fungus on leaves makes diagnosing powdery mildew easy. In northern states, powdery mildew typically appears in mid- to late summer. In severe instances, if left untreated, powdery mildew can cause leaves to turn yellow, die and fall off. In most instances, it is only an unsightly nuisance. To suppress powdery mildew, improve air circulation by thinning and properly spacing plants. Avoid fertilizers that are high in nitrogen, since they promote new growth that is especially susceptible to powdery mildew. Control may also be achieved with the use of fungicides applied as soon as symptoms are visible.

Downy mildew, caused by the fungus-like organism *Plasmopara halstedii*, causes yellow to dark blotchy areas on upper leaf surfaces and grayish white fuzzy growth on leaf undersides. This disease is becoming an increasingly important problem in nurseries. During cool, wet weather, downy mildew can cause severe blight on variety 'Goldsturm,' but a number of other rudbeckia varieties are less susceptible. Symptoms may appear in both spring and fall. Cultural practices to reduce humidity and the duration of leaf wetness periods, such as spacing plants to encourage air movement and watering the soil rather than the foliage, can help to slow the progress of the disease. Sprays of labeled fungicides, beginning as soon as symptoms appear, can also suppress downy mildew.

Several species of *Uromyces, Puccinia, and Aecidium* rust fungi produce powdery, rusty red spores in blister-like pustules on leaves and stems. Some of these rusts can be partially controlled by removing nearby plants that act as their alternate hosts; labeled fungicides are also used to suppress rusts.

Symptoms of stem rot, caused by the soilborne fungi *Sclerotium rolfsii* and *Sclerotinia sclerotiorum*, include yellowing of lower leaves followed by wilting and death of the entire plant. Both fungi produce a mass of cottony white mycelium at the soil line; *S. rolfsii* produces abundant pinhead-size, white to reddish brown sclerotia in the mycelium, whereas *S. sclerotiorum* sclerotia are coal black, larger, mouse-dropping-size, and often imbedded in the center of the lower stem. Control involves removing and destroying all infected plants, and even removing and replacing the sclerotia-infested topsoil around the plant. Several fungicides are labeled as drench treatments for control of these stem rots, but should be used preventively. Verticillium wilt, caused by the soilborne fungus *Verticillium dahliae*, colonizes the vascular system and blocks water flow, resulting in yellowing of leaves followed by browning and wilting. Cutting open the stem near the base will reveal browning of the vascular tissues. Diseased plants should be removed and destroyed. Since the fungus survives for many years in soil, replant infested areas with plants that are not hosts of Verticillium wilt. *Fusarium oxysporum* has also been associated with wilting rudbeckias.

Bright yellow mottling is a sign of *Tobacco streak virus* infection on rudbeckia.

Aster yellows causes greening of rudbeckias.

Bacterial leaf spot caused by *Pseudomonas cichorii* produces angular water-soaked spots on rudbeckia.

Downy mildew on rudbeckia has emerged as a serious problem in late summer in many parts of the country.

Aster yellows causes eye-catching deformities in the normal growth pattern of rudbeckia: leaves that emerge from the center of flowers, flowers that sprout out of the seed head, witches' brooms, yellowing, and dwarfing. The pathogen, a phytoplasma, is spread from plant to plant by leafhoppers. Infected plants should be removed and destroyed. Early-season control of leafhoppers by insecticide sprays and removal of weed hosts may help minimize disease spread.

Virus diseases on rudbeckia include rudbeckia mosaic (RuMV), potato yellow dwarf (PYDV), tomato spotted wilt (TSWV), tobacco streak (TSV), and Bidens mottle (BiMoV). Symptoms include mottled yellow-green discoloration of leaves, distortion, dwarfing, and even browning of leaves. Plants confirmed as having virus infections should be removed and destroyed. Confirming the identity of a virus requires a lab test, but is useful because each virus is spread in characteristic ways (for example, by certain insects, by handling, or by seed), and has its own host range. Once the virus is identified, a plan to prevent its spread can be developed.

The declining rudbeckia is infected with Fusarium wilt.

Septoria leaf spot on rudbeckia is typified by reddish, angular spots that turn brown over time.

Powdery mildew appears on upper leaves of rudbeckia in round patches.

Pseudomonas and *Xanthomonas* bacteria can blight leaves (left) and stems (right) of rudbeckia under wet conditions.

Ruta

Rutaceae

Ruta, or rue, includes herbs with ornamental qualities, particularly the species *Ruta graveolens*, common rue, which is a sub-shrub with blue foliage. Rue should be grown in full sun or partial shade, in very well-drained soil. It grows well in hot, dry conditions and will not tolerate waterlogged soil. No diseases have been reported for *Ruta* from North America.

Salvia

Lamiaceae

Salvia, or sage, is a large genus of often-aromatic herbs and ornamentals, many of which are only half-hardy north of Zone 8. *S. officinalis* is the sage used to flavor turkey stuffing; it and its varieties also have numerous ornamental uses. *Salvia splendens*, scarlet sage, although often grown as an annual, is actually a tender tropical perennial. Grow salvias in either full sun or part shade, depending on the species. Moist sites with excellent drainage and abundant organic matter are ideal for most salvias. Plants with hairy leaves should be given particularly well-drained sites, in full sun. *Salvia greggii*, Texas sage, and some of its hybrids are drought, heat and humidity tolerant. Wet winter conditions can be lethal to salvias.

Plants may be susceptible to powdery mildew, which results in patches of white, powdery-looking fungal growth on foliage. Symptoms can be troublesome during greenhouse production, but often appear late in the growing season in the landscape. In greenhouses, reducing humidity, increasing air flow at the plant level, and spraying labeled fungicides can help to suppress powdery mildew.

Downy mildew of salvia is caused by species of *Peronospora*. Leaves may become mottled and yellowed, so that the plant looks nutritionally deficient. Angular yellow blotches can be seen between the leaf veins. Downy mildew symptoms may also resemble injury from foliar nematodes; both cause angular lesions. However, leaves with downy mildew will show fluffy grayish brown to purple growth on the underside of leaves. When possible keep greenhouse relative humidity below 85 percent.

Fungal leaf spot diseases of salvia are caused by species of *Cercospora, Ramularia, Corynespora, Alternaria*, and *Entyloma*, as well as *Myrothecium roridum*. Corynespora leaf spot symptoms include tiny, black, irregularly shaped spots on lower leaves, or black, sunken areas on lower portions of the stems. Stem lesions are elongated and cause girdling and plant death on some varieties. Alternaria leaf spot begins as small water-soaked areas that soon become reddish brown or black circles about 1/8 inch across. To reduce the risk of infection by leaf spot fungi, it is helpful to keep the foliage dry by using drip irrigation rather than overhead watering, and by spacing plants to encourage air flow and rapid drying. White smut, caused by *Entyloma compositarum,* results in round, white to tan leaf spots, up to ½ inch across, with indistinct margins. The spots later turn brown and necrotic, followed by death of the entire leaf. Several fungicides are labeled for control of leaf spot fungi on *Salvia*. Bacterial leaf spot diseases that occasionally attack salvia are caused by *Pseudomonas syringae* pv. *maculicola* and *Pseudomonas cichorii*. To lessen the risk of bacterial leaf spot, avoid overhead watering and do not crowd plants. Aster yellows, caused by a phytoplasma, can deform plants, resulting in shoot proliferation, leaves developing from flowers, and other abnormalities. Infected plants should be pulled out and destroyed. Several species of *Puccinia, Uromyces, and Aecidum* rust fungi produce small, cinnamon- to dark brown-colored, blister-like spots called pustules on leaves. Heavily infected leaves may turn yellow but rust diseases seldom kill plants. Increasing air flow in plantings and using drip rather than overhead irrigation can reduce the risk of rusts. Preventive fungicide sprays can also be helpful.

Fungi that cause stem rots of salvia include *Sclerotium rolfsii* and *Sclerotinia sclerotiorum*. These fungi attack at the soil line, resulting in yellowing and death of stems and entire plants along with white masses of fungal growth. Sclerotia of *S. rolfsii* are numerous, white to reddish-brown, about pinhead-size, and appear in the fungal growth at the base of the diseased stem. Sclerotia of *S. sclerotiorum*, on the other hand, are black, look like mouse droppings, and are frequently found inside collapsed stems. Infected plants should be dug up and destroyed, taking care not to scatter the sclerotia. For highly localized outbreaks, surface soil surrounding the diseased plants can be removed and replaced with non-infested soil. Drenches with labeled

Corynespora leaf spot on sages can be either dark brown or black target spots.

Myrothecium leaf spot can be a problem on sage especially during propagation of cuttings.

Pseudomonas leaf spot is expressed as angular spots on sage in propagation as well as in the landscape.

Downy mildew on sage is common in production and the landscape across the United States.

fungicides can suppress *S. rolfsii* and *S. sclerotiorum*, but avoidance is the most effective strategy. An additional stem rot fungus on *Salvia* is *Phymatotrichopsis omnivora*, another soil native with a very broad host range. Fortunately, it is restricted geographically to a few regions of the extreme southern United States. The only effective means of disease control is to avoid infested areas. Another soil-dwelling fungus, *Verticillium albo-atrum*, occasionally causes wilting and death of salvia. The fungus invades through the roots and blocks the water-conducting tissues of the plant, resulting in wilitng. Infected plants should be removed and destroyed; avoid planting other species that are hosts of this fungus in the same locations from which the sick plants were removed.

Damping-off, caused by *Pythium* and *Rhizoctonia* species, attacks germinating seeds and young seedlings, resulting in softening (*Pythium*) or withering (*Rhizoctonia*) of the lower stem or hypocotyl, followed by seedling death. *Pythium* species are active when the soil or growing medium remains waterlogged for long periods, whereas *Rhizoctonia solani* attacks when the seedlings are under stress from unfavorable temperatures, dryness, nutrient levels, or other environmental conditions. To minimize damping-off, it is helpful to maintain conditions that favor rapid germination and seedling development. Fungicide soil drenches are also used to suppress damping-off. Salvias may be attacked by these same damping-off fungi when they are mature plants, with stem rot or root rot as a result.

Powdery mildew on sage causes characteristic white patches.

Phytophthora root and crown rot on salvia is caused by species of *Phytophthora*. The crown and lower stem of plants attacked by these water molds may show a dry rot at or near the soil line. Since the disease can be spread via diseased plants and infested soil, suspect plants should be discarded and destroyed. Fungicide drenches are effective when used preventively. Since *Phytophthora* species are most active in heavy clay soils, it may help to incorporate adequate amounts of soil amendments such as compost or peat. Raised beds can improve drainage, and so can reduce the risk of Phytophthora root and crown rot. Irrigate deeply and infrequently, allowing the top inch of soil to dry out before irrigating.

Nematodes that attack *Salvia* spp. include root-knot nematodes (*Meloidogyne*) and the foliar nematode *Aphelenchoides fragariae*. Root-knot nematodes attack root systems from the soil, causing the swellings that give the disease the name "root knot"; in severe cases, the root system is grossly deformed, and functions poorly. Above-ground symptoms may include yellowing and stunting. Rotating to plants that are not hosts of *Meloidogyne* spp. for several years will reduce root-knot risk. Some varieties of salvia have also been shown to be resistant to certain species of *Meloidogyne*. Foliar nematodes cause brown areas on leaves, often limited by leaf veins so they are angular in shape.

Among the viruses that affect salvia are tomato spotted wilt (TSWV), potato yellow dwarf (PYDV), and cucumber mosaic (CMV). Symptoms can include mottled discoloration of foliage and stunting. Virus-infected plants should be removed and destroyed to reduce spread of the problem. It is helpful to have the specific virus present diagnosed by a plant clinic; since each virus is spread in different ways, knowing the identity of the virus helps point the way to strategies that control its spread.

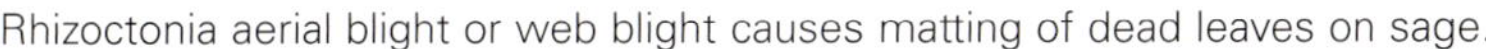

Rhizoctonia aerial blight or web blight causes matting of dead leaves on sage.

The black sclerotia of *Sclerotinia sclerotiorum* can develop in or on sage stems and crowns.

Santolina

Asteraceae

Santolina chamaecyparissus, lavender cotton, is hardy from Zones 6-8, and *Santolina rosmarinifolia*, green lavender cotton, is slightly less hardy. They are common as edgers. Plants can tolerate summer heat, but not heat coupled with humidity as in the American South; they develop empty centers in this climate.

The only leaf spot disease that has been reported from santolina in North America is caused by the fungus *Alternaria tenuissima*. To combat fungal leaf spots, avoid overhead watering and space plants to encourage air flow and rapid drying. Several fungicides are labeled against fungal leaf spots.

Santolina is also susceptible to fungal root rots caused by *Rhizoctonia solani* and *Fusarium oxysporum*, plus Pythium and Phytophthora root rots. These soil-borne problems can cause yellowing, wilting, and plant death. To combat Pythium and Phytophthora root rot, select well-drained sites, improve drainage by using raised beds, and avoid over-watering. Fungicides can be used as soil drenches to suppress these diseases during nursery production. Infected plants should be removed and destroyed.

Charcoal rot, caused by the fungus *Macrophomina phaseolina*, causes yellowing and collapse of plants. Splitting open the stem at the base will reveal abundant, ground-pepper-size, black particles; these are sclerotia of *M. phaseolina*. Infected plants should be removed and destroyed.

One of the few enemies of santolina is winter injury. Prune out dead areas in spring to make way for healthy new foliage.

Saxifraga

Saxifragaceae

Saxifraga, or saxifrage, is a large genus of plants that are primarily appropriate for rock gardens, as they require excellent drainage. Cooler climates are ideal and partial shade is needed to protect plants from sun. *Saxifraga stolonifera*, also known as strawberry begonia and strawberry geranium, is one of the most familiar of the saxifrages because it is a garden plant in the South and a houseplant in the north. Late frosts may damage some of the early-flowering species, and winter moisture may be harmful. Some mossy carpet-forming saxifrages will melt out in the center if humidity is too high.

Several rust fungi infect saxifrage, including *Puccinia* and *Melampsora*. Medium to dark brown pustules appear on the bottom of the leaf; when spots are older, they often show as chlorotic spots on the upper surface of the leaf. Leaf pustules contain spores that can spread by air movement or splashing water. Rust can be introduced on infected plant material, so scout incoming plants for symptoms. Good air circulation is essential. Avoid overhead watering and space plants adequately to enhance drying after rains. Several fungicides are labeled against rust.

A few leaf spot fungi—*Cercosporella, Phyllosticta, Septoria,* and *Ramularia*—can attack saxifrages, but seldom cause serious damage. The same practices used for managing rust fungi will help to control fungal leaf spot problems. Note that saxifrages are also susceptible to foliar nematodes (*Aphelenchoides*); although the symptoms may be similar to fungal leaf spot symptoms, fungicides will be of no use. Foliar nematodes cause vein-limited patches of discoloration in the foliage. The nematode culprits themselves may be viewed if discolored tissue is torn into pieces in a dish of water and examined 1 to 24 hours later with a strong hand lens or microscope.

Powdery mildew may coat the leaves with a white powdery growth. Gray mold blight, caused by *Botrytis cinerea,* results in browned leaves and stems, and growth of a fuzzy layer of grayish-brown fungus on infected plant parts. The same practices used against rust will help to minimize the risk of powdery mildew and gray mold outbreaks.

The only virus disease reported from saxifrages is impatiens necrotic spot (INSV). Symptoms are variable but can include yellow mottling of leaves and stunting. To combat INSV during greenhouse or nursery production, it is necessary to regularly scout incoming stock and production areas for thrips, the insects that spread INSV, and to watch for symptoms on the many plant species that are susceptible. Control thrips with insecticides if they are found. Infected plants should be removed and destroyed.

Foliar nematode infestation on saxifrage appears as blotchy dead spots within the leaf.

Cercosporella leaf spot on saxifrage.

Scabiosa

Dipsacaceae

Scabiosa, or pincushion flower, is best suited for areas that do not have high heat and humidity. Scabiosas grow best in full sun and soils should be well-drained and rich, with a neutral or slightly alkaline pH. Regular deadheading keeps scabiosas flowering for as long as possible. Division every three years also helps to rejuvenate the planting. The pincushion flower, or scabious, *S. caucasica*, is widely grown and has many varieties. Recently, the more compact *S. columbaria* has become very popular: it includes the cultivar 'Butterfly Blue'.

Plants with powdery mildew will have patches of white, powdery-looking fungal growth on foliage and stems. Symptoms can appear in spring during nursery production, but often appear later in the growing season in the landscape. During production, reducing humidity, increasing air flow at the plant level, and spraying fungicides preventively can help to suppress powdery mildew. If control is needed in the landscape, increasing plant spacing and avoiding deep shade are cultural controls, and fungicides can also be used.

Sclerotium rolfsii and *Sclerotinia sclerotiorum* attack at the soil line and lower stem, resulting in yellowing and death of stems and entire plants, accompanied by white masses of fungal growth that appear on the lower stem. Sclerotia of *S. rolfsii* are numerous, white to reddish-brown, about pinhead-size, and appear in the fungal growth at the base of the diseased stem. Sclerotia of *S. sclerotiorum*, on the other hand, are black, somewhat cigar-shaped, up to half an inch long, and are frequently found inside the stem base. Infected plants should be dug up and destroyed, taking care not to scatter the sclerotia. Another stem rot fungus on scabiosa is *Phymatotrichopsis omnivora,* which has a very broad plant host range but is restricted geographically to a few areas of the southwestern United States. It spreads slowly from plant to plant when a fungal strand from an infected root grows through the soil to a nearby healthy root. The only effective means of disease control is to avoid infested areas. Other root and stem rots found on *Scabiosa* spp. are caused by species of *Rhizoctonia*, *Pythium*, and *Thielaviopsis*. To minimize the risk of attack by the fungi *Rhizoctonia* and *Thielaviopsis*, it helps to avoid prolonged exposure to stressful conditions such as nutrient imbalances or cold soils. Pythium root rots can be suppressed by preventing prolonged periods of waterlogged soils. Fungicide drenches can also help against these pathogens.

Aster yellows, caused by a phytoplasma, can cause growth deformations such as shoot proliferation and leaves emerging from flowers. The phytoplasma is spread by leafhoppers. Infected plants should be removed and destroyed. Insecticides can be used to suppress leafhopper populations if the situation warrants.

Virus diseases reported from scabiosa include tomato spotted wilt (TSWV) and beet curly top (BCTV). Symptoms can include mottling of foliage and stunting of plants. Plants showing virus symptoms should be removed and destroyed. It is helpful to have the problem virus identified by a diagnostic clinic; diagnosis can point out which insects are important to control to prevent spread of the virus, and which additional plant species are in the host range of the identified virus.

Aster yellows causes infected scabiosas to develop green flowers.

Scabiosa leaves with downy mildew develop angular dead spots.

Powdery mildew on scabiosa is sometimes seen as purple discoloration, while at other times there are more typical white frosty patches of fungal growth.

Scutellaria

Lamiaceae

Scutellaria or skullcap is a large genus with only a few garden plants. These are grown in full sun in the North, while some afternoon shade is helpful in the South. Well-drained, moderately fertile soils are needed for the two most common garden species, the upright *S. altissima* and the low, spreading *S. alpina*. Scutellarias may benefit from the addition of lime if soils are acid.

Several fungal leaf spots, caused by species of *Cercospora*, *Septoria*, and *Phyllosticta*, are the most widespread disease problems of scutellaria. Symptoms are dark spots of varying sizes on leaves; leaves may wither and drop off if spots are numerous. To reduce the risk of fungal leaf spots, use drip irrigation and space plants to encourage air flow and rapid leaf drying. Fungicides can be used to supplement these cultural controls.

A stem rot of scutellaria can be caused by the fungus *Botrytis cinerea*. Sunken, dark brown areas at the base of the stem can be accompanied by browning and collapse above the affected point. Under humid conditions, a fuzz of grayish-brown fungal growth can develop on the diseased part of the stem. To control the risk of *B. cinerea* in greenhouses, heating, venting, and air flow should be used to keep relative humidity below 85%. Fungicides can be used to provide an added measure of protection.

Powdery mildew of *Scutellaria* spp. produces patches of white, powdery-looking fungal growth on foliage. Symptoms can appear in spring during nursery production, but often appear late in the growing season in the landscape. During production, spacing plants adequately and spraying fungicides can reduce powdery mildew severity. In the landscape, increasing plant spacing and avoiding deep shade are cultural controls; fungicides can also be used.

Root rots caused by *Phymatotrichopsis omnivora* and *Rhizoctonia solani* can induce yellowing, wilting, collapse of foliage, and stunting of plants in less severe cases. A soil native, *P. omnivora* has a very broad host range but is restricted geographically to a few areas of the extreme southern United States. It spreads slowly from plant to plant when a fungal strand from an infected root grows through the soil to a nearby healthy root. The only effective means of disease control is to avoid infested areas. To minimize the risk of attack by *Rhizoctonia solani,* avoid prolonged exposure to stressful conditions such as nutrient imbalances, excessively dry soils, or cold soils. Fungicide drenches can also help suppress Rhizoctonia root rot.

The only virus reported from scutellaria in North America is *Alternanthera mosaic virus* (AltMV). Leaf symptoms include yellowish mottling, yellow to brown concentric rings, and yellow patterns of wavy lines. Plants with symptoms should be removed and destroyed to minimize spread of the disease.

Powdery mildew on scutellaria shows typical white patches.

Sedum

Crassulaceae

Sedum, or stonecrop, includes primarily ground-hugging succulent plants, used largely for groundcovers and in rock gardens but with a few popular upright clumping forms as well. Neutral to alkaline soils low in fertility work well for sedums, and good drainage is required. Sedums should be grown in full sun and watered only as needed. Divide upright clumping forms every 3 or 4 years to maintain vigor.

Nursery production of sedums presents a challenge, as overfertilization and overwatering will result in collapse of stems from bacterial and fungal diseases. Stem rots can be caused by various soilborne microorganisms, including several fungi (*Sclerotium rolfsii, Rhizoctonia solani,* and *Fusarium solani*) and the fungus-like organism *Phytophthora* sp. Yellowing of lower leaves, wilting and death of the plant may be due to *S. rolfsii.* A white cottony mass of fungus growing on the crown or on the soil near the crown distinguishes this rot from others. In this fungal growth may be found numerous, whitish to reddish-brown, spherical sclerotia that are about the size of pinheads. Control can include removing and destroying all infected plant parts, and even removing topsoil around the plant and replacing it with new soil. *Rhizoctonia solani* and *Fusarium solani* can rot both crowns and roots. Sunken lesions may occur at the soil line. Remove and destroy infected plants to slow spread of these diseases. Drench treatments with fungicides may suppress fungal stem and root rots during nursery production.

Anthracnose is caused by the fungus *Colletotrichum trifolii.* On popular cultivars such as 'Autumn Joy' and 'Frosty Morn' it produces a range of damage from a few irregular, blackened areas on resistant stems to large, sunken, oval to diamond-shaped lesions on stems of susceptible plants. The lesions are straw colored with brown borders. Tiny black fruiting structures called acervuli develop in the bleached areas and are readily visible with a hand lens. As the spots enlarge, they grow together and can girdle and kill one or more stems on a plant. Start with healthy transplants in a well-drained soil, provide balanced nutrition, and avoid over-watering.

Gray mold flower and leaf blight is caused by the fungus *Botrytis cinerea*. This fungus occurs everywhere and commonly infects senescing or damaged plant parts such as old flowers, resulting in a fuzzy gray mold. Sedum cuttings may develop a dark brown rot at the cut end. Disease is favored by cool, wet conditions and the presence of withered leaves or old flower petals. Sanitation is the most important means of disease prevention but control may also be achieved with the use of labeled fungicides applied as soon as symptoms are visible.

Rust, caused by the fungi *Puccinia rydbergii* and *P. umbilici*, produces blister-like pustules on leaves that break open to release brick red, powdery spores. Leaves on heavily infected plants may turn yellow, and growth may be stunted. Fungicides need only be applied at the onset of symptoms.

Several fungi, including *Cercospora, Corynespora, Septoria, Pleospora, Phyllosticta*, and *Stemphylium* spp., can cause fungal leaf spots on sedum. Leaf spots are typically sharply defined, more or less rounded areas of discoloration; in severe cases, leaves may turn yellow and wither. Fungal leaf spots are favored by prolonged wet conditions. Cultural management includes watering at the soil level rather than overhead, and spacing plants far enough apart so that they dry off rapidly. Under weather conditions highly favorable for disease, fungicides can be applied, starting as soon as the first spots appear.

Soil-borne fungi such as *Fusarium* and *Pythium* spp. cause collapse of container-grown sedum.

Powdery mildew on sedum is quite common in nursery production, and is becoming more common in the landscape. Leaves infected with powdery mildew show brown scab-like spots that develop with very little if any of the white powdery growth that is typical on other hosts. Look closely with a hand lens for the fine, white fungal threads and chains of powdery mildew spores. Cut plants off at the soil line in the fall and bury or compost the debris, in order to reduce the amount of fungus that survives the winter. Watering should be done early in the day so plants can dry before nightfall. Plants that are overcrowded or growing in shaded areas may be prone to powdery mildew. Sometimes the infection appears on the undersurfaces of the leaves. Many plants develop powdery mildew late in the season and do not need treatment with fungicides. Fungicides applied after the symptoms have appeared will protect new growth, but will not make the scabby lesions disappear. Wettable powder formulations are advisable for use on sedum, to avoid plant injury that may result from the use of emulsifiable concentrates.

A soft rot of sedum stems is caused by the bacterium *Pectobacterium carotovorum*. Stems become darkened and soft at the base, and a foul smell develops from the diseased tissue. Diseased plants should be removed and destroyed. Root knot nematodes (*Meloidogyne*) cause rounded swellings on roots, and can severely deform the root system in some cases. Aboveground growth can be yellowed and stunted. Affected plants should be removed and destroyed. *Tomato spotted wilt virus* (TSWV) and *Impatiens necrotic spot virus* (INSV) can cause a range of symptoms on foliage including ringspots and mottling. These viruses are spread in greenhouses and in the field by thrips. Sources of the virus may include a wide range of other crops and certain weeds. *Arabis mosaic virus* (ArMV), a nepovirus, has been reported in sedum from Israel. Virus management techniques for nurseries include rigorous scouting of incoming plant materials for symptoms in order to catch severe problems before the plants are moved into production areas. Scout for thrips in nurseries and greenhouses, and apply insecticides as needed. Controlling weeds in production facilities is also important.

Purplish and necrotic ringspots in sedum leaves indicate a virus infection.

Pectobacterium sp. causes a mushy, wet rot on infected sedum.

Powdery mildew can appear as white or scabby tan to gray patches (left) or as black diffuse spots (right) on infected sedum.

Sempervivum

Crassulaceae

Sempervivum, or hens and chicks, is a genus of succulents used for rock gardens and edging with full sun exposure. Many varieties are available for *Sempervivum tectorum*, the common houseleek. Soil of low fertility that is very well drained is optimal and especially important in the South. Plants are not as drought tolerant as they appear and plants need to be protected from wetness in winter. Each rosette will die naturally after flowering.

A rust fungus, *Endophyllum sempervivi*, produces striking symptoms on hen and chicks: infected leaves tend to grow several times their normal length, since the infection changes the normal flow of growth hormones. These elongated leaves eventually develop small, cream-colored to yellowish blisters that split open to release powdery, orange spores. Remove all affected rosettes to reduce disease spread. Sprays of fungicides labeled against rusts may help to suppress the disease.

Rhizoctonia solani can cause an aerial blight of sempervivums under very warm and humid conditions. Lower stems and leaves may darken and soften, and the diseased area and nearby soil may be covered with a web of grayish fungal strands. Under favorable environmental conditions, the blight can start from a leaf and move quickly to the whole rosette, then other rosettes. Control starts with modifying the environment to minimize high humidity around the plants. Keep mulch away from the leaves to

Drought on hen and chicks can result in loss of the plants along the outer margin of the colony.

allow good air movement, and irrigate in late morning or at midday, no more frequently than required. Fungicides can also be used to suppress disease development.

Anthracnose, caused by the fungus *Colletotrichum gloeosporioides*, has been reported from nursery production facilities in Florida. Symptoms start as a blackened rot of leaves that may either remain localized or spread to the rest of the plant. Plants showing symptoms should be removed and destroyed. Fungicides can be used to suppress the disease.

The fungus-like organisms *Phytophthora nicotianae* and *P. drechsleri* can cause a leaf and stem rot of hens and chicks when cool winter or spring temperatures are accompanied by damp conditions. Older leaves become dull and yellowish, then translucent, with brownish tips. Leaf and stem rot may spread from older to younger leaves, occasionally killing plants. Affected leaves should be removed promptly. A related organism, *Pythium*, can cause root rots under cool conditions in persistently wet soil. Reducing humidity, raising greenhouse temperature, and avoiding overly wet soil may help against both diseases. Several fungicides are effective against *Pythium* and *Phytophthora* spp.

Silene

Caryophyllaceae

Silene, or campion, includes garden species, some of them suitable for rock gardens. *Silene acaulis*, moss campion, forms cushions that fare well if drainage is extremely good. *S. virginica*, native to the southeastern United States, also enjoys soils that are well drained. Campions are best grown in full sun.

Rusts, the most widely reported diseases of campion, are caused by *Uromyces silenes, Uromyces suksdorffii,* and *Puccinia aristidae*. Small, faint yellow spots on leaves develop blisters (primarily on the undersurfaces of the leaves) that ultimately break open to release powdery, brick-red spores. Severely infected leaves may turn yellow and wither. Cultural controls include avoiding overhead watering and spacing plants widely enough to insure rapid drying. Remove and destroy heavily infected leaves and plants. Fungicides are also available for plant protection. Fungal leaf spots can be caused by species of *Ascochyta, Marssonina, Phyllosticta,* and *Septoria*. Rounded leaf spots can cause leaves to turn brown and wither if they are numerous. Keep foliage as dry as possible by minimizing overhead watering and selecting sites with good air drainage. Remove and destroy heavily infected leaves and plants. Fungicides are also an option as a supplement to cultural practices.

Damping off of seedlings in production nurseries can be caused by the fungus *Rhizoctonia solani*. Seedlings develop a dark, constricted area at the soil line and collapse. Cultural controls begin with keeping temperature and media moisture within ranges favorable to the plant. Fungicide drenches can supplement cultural controls.

Flower smut is caused by the fungus *Sorosporium saponariae*, and anther smut by the fungus *Microbotryum violaceum*. Flower parts become blighted and produce masses of powdery black spores. Heavily infected plants should be removed and destroyed. Fungicides are an option for control.

Downy mildew is caused by the fungus-like organism *Peronospora silenes*. Leaves can develop yellowed patches, followed by browning and collapse. Prolonged wet conditions favor downy mildew. Cultural controls are the same as for fungal leaf spots. Some of the fungicides labeled for downy mildew are different from those labeled for fungal leaf spots.

Carnation etched ring virus (CERV) is the only virus problem reported from campion. Brown lines and blotches develop on foliage. Infected plants should be removed and destroyed to reduce transmission of the virus by aphids. In plant production situations, scout for aphids and apply insecticides as indicated by the scouting results.

Solidago

Asteraceae

Solidago, or goldenrod, is better appreciated for its garden contributions in Europe than in North America, where many gardeners still perceive it as a weed. *Solidago* has recently acquired new cachet as a native plant, however. Hybrid varieties are available; *Solidaster* (x*Solidaster luteus*) is thought to be a natural interspecific hybrid of *Solidago* and *Aster ptarmicoides*. *Solidaster* appears to be vulnerable to some of the same diseases as its parents. Both solidasters and goldenrods require sunny locations and moderately fertile, well-drained soils, and benefit from good air movement to reduce rust and powdery mildew.

Rusts and powdery mildews are the most common diseases of goldenrods. Rusts, caused by species of *Coleosporium*, *Puccinia*, and *Uromyces*, appear as small, bright orange, yellow, or brown pustules on the underside of leaves. Plant resistant varieties—for example, 'Fireworks' and 'Golden Fleece' are resistant to both rusts and powdery mildews—and provide maximum air circulation. Clean up all debris, avoid overhead watering, and water only during the day so that plants will dry quickly. Several fungicides labeled for rust can be used as needed. Powdery mildew, caused by several different fungal species, is usually found on goldenrods that do not have enough air circulation or adequate light. Problems are worse where nights are cool and days are warm and humid, resulting in dew formation overnight. The powdery white or gray fungus is most evident on the upper surface of leaves. Leaves will often turn yellow or brown, curl up, and drop off; new foliage emerges crinkled and distorted. Use the same methods that are described above for the control of rusts.

Leaf-spotting fungi on goldenrod include species of *Ascochyta*, *Asteroma*, *Cercospora*, *Colletotrichum*, *Macrophoma*, *Phyllosticta*, *Placosphaeria*, *Ramularia*, and *Septoria*. Small, dark spots appear first on lower leaves. Leaves with numerous spots may turn yellow, then brown, and wither. Avoid overhead watering, space plants adequately to insure rapid drying, and water early in the day. Fungicides are rarely needed for leaf spot management on goldenrod. A bacterial leaf blight caused by a *Xanthomonas* species occurs on *Solidago gigantea*. In general, the same cultural controls as for fungal leaf spots will be helpful in suppressing bacterial leaf blight. Copper-containing bactericide/fungicides are also an option as needed.

Aster yellows, caused by a phytoplasma and spread by leafhoppers, can cause plants to be chlorotic and flowers to develop abnormalities, including absence of their usual pigmentation. Affected plants should be removed.

Crown gall, caused by *Agrobacterium tumefaciens*, can occur on goldenrod when infected cuttings are used for propagation. The galls start on the wounded cut ends and slowly enlarge up to an inch or so in diameter. They appear right at the soil line on stems or sometimes on wounded parts

Aster yellows on goldenrod causes bunching of foliage as well as abnormal flowers.

above that area. The bacterium can become a resident of soil and infect healthy plants as well. The pathogen has a very wide host range, so other perennials may become infected if planted into infested soil. Lesion nematodes (*Pratylenchus pratensis*) can damage root systems of goldenrod with their underground feeding, resulting in stubby roots with darkened, decayed portions. Remove and destroy infected plants. Nematicides are an option in nursery production areas.

Plants infected with *Tomato spotted wilt virus* (TSWV) may show a range of symptoms including yellow or brown rings on leaves, blighting or dwarfing of shoots, and a mottled pattern of leaf yellowing. Infected plants should be removed and destroyed promptly. In greenhouses, monitoring for the insects that spread TSWV (western flower thrips) can guide the timing of insecticide sprays.

Rust often appears on goldenrod as small chlorotic spots on the upper leaf surface (left) and later develops characteristic orange sporulation (right).

Stachys

Lamiaceae

Stachys, or betony, includes some rhizomatous or stoloniferous perennials that are popular in the garden. *Stachys byzantina*, lamb's ears, is the best known plant, renowned for its woolly white leaves. Some varieties do not produce flowers at all and are propagated vegetatively. Stachys are used as ground covers, edging plants, or in rock gardens. They are grown in full sun, in moist, reasonably fertile, well-drained soil. Hot, humid weather causes foliage to deteriorate, especially when combined with overhead watering or summer rainfall. Some afternoon shade helps plants to survive southern climates.

Powdery mildew is the most common disease problem of stachys. Patches of granular-looking, white to grayish fungal growth appear on leaves. In severe cases, the patches can grow together and nearly cover the leaf surface. Avoid overhead watering, and water in the morning to allow beds to dry before evening. Space plants widely enough to encourage good air movement and rapid drying. Fungicides can supplement good cultural practices but since stachys has hairy leaves, test sprays on a small scale to be sure of plant safety under your conditions. Rust, caused by a *Puccinia* species, is relatively uncommon on stachys. Rust causes blister-like pustules on

Phytophthora crown rot can cause a rapid death at the end of summer on stachys.

leaves that break open to release powdery, reddish brown spores. Fungal leaf spot diseases are caused by *Septoria, Cercospora, Cylindrosporium, Ovularia, Phyllosticta,* and *Ramularia*. Leaf spot fungi cause tan to dark brown spots on leaves, often beginning on the lower leaves. Numerous spots may cause a leaf to turn yellow or brown and wither. Use the same controls as described above for powdery mildew.

Phytophthora root rot is caused by *Phytophthora* species. A dark discoloration and softening of roots and the base of the stem lead to collapse and death of the plant. Select well-drained sites or container growing media, or use raised beds in poorly drained soil and avoid overwatering. Fungicide drench treatments may suppress the disease, but should be used in tandem with cultural controls.

Root knot nematode, *Meloidogyne*, can cause swellings and deformation of the root system. Foliage may become stunted and yellowed as functioning of the root system declines. Remove and destroy infected plants, and replant with non-hosts of root knot nematode.

Powdery mildew colonies are sometimes hard to see on the hairy white leaves of stachys.

Stokesia

Asteraceae

Stokesia, or Stokes' aster, is native to the southern United States, where it is evergreen. The genus contains only one species, *S. laevis*, whose varieties should be grown in sun or light shade in light, well-drained, acidic soil. Avoid planting in heavy soils that will keep plants wet during the winter. Mulch to protect stokesia in colder climates. Flowering will be promoted by deadheading.

Fungal diseases that can mar stokesia include leaf spots caused by *Ascochyta*, *Alternaria*, and *Colletotrichum*, a head blight of flowers caused by *Botrytis cinerea*, and aerial blight caused by *Rhizoctonia solani*. The fungal leaf spots typically appear first on lower leaves, as tan to dark brown spots; when they are numerous, a leaf can turn yellow or brown and wither. Head blight appears as a

Spotting on Stokes' aster due to *Tomato spotted wilt virus* (TSWV).

fuzzy grayish mass of fungus on flowers, and Rhizoctonia aerial blight shows up as webs of grayish fungal growth on the stems and leaves, particularly in greenhouses. All of these diseases are favored by prolonged periods of wet conditions. To suppress them, avoid overhead watering, water in the morning, and space plants far enough apart to allow rapid drying. Fungicides are available as needed to supplement cultural control methods.

Powdery mildew forms patches of grayish white, granular-appearing fungal growth on leaf surfaces. Cultural controls, including selecting sites with good air movement and spacing plants far enough apart to allow for rapid drying, can be supplemented by fungicide sprays.

Bidens mottle virus (BiMoV) and *Cucumber mosaic virus* (CMV) have been confirmed from stokesia.

Alternaria leaf spot on Stokes' aster first appears as small black speckles.

Rhizoctonia solani can cause leaf spots as well as web blight on Stokes' aster.

Colletotrichum sp. causes anthracnose on Stokes' aster.

Tanacetum

Asteraceae

Tanacetum, or tansy, is a chrysanthemum relative that does well in fairly dry conditions in sunny sites. Afternoon shade is acceptable. This genus now includes costmary (*T. balsamita*) and painted daisies and pyrethrums (*T. coccineum*), as well as feverfew (*T. parthenium*). Common tansy, *T. vulgare*, is one of the easiest to grow, and is on rare occasions used medicinally: all herbs in this genus are very toxic. Common tansy has very aromatic foliage and yellow flower heads that lack ray flowers, so they look button-like. Tansies are best grown in light soils but will tolerate all but the heaviest of soils. *Tanacetum* spp. may be invasive in the garden: feverfew, in particular, will self-seed abundantly.

Powdery mildew appears as white, granular patches on leaves. In severe cases, it can nearly coat the leaf surfaces, and turn leaves yellow. Powdery mildew is favored by prolonged periods of high humidity and stagnant air. To suppress it in greenhouses, heat and ventilate to minimize high-humidity periods. Outdoors, select sunny sites with good air movement, and avoid crowding plants. Fungicides can suppress the disease.

Bacterial fasciation, caused by the bacterium *Rhodococcus fascians,* appears in greenhouse production as an abnormal proliferation of small shoots from the base of the plant. Other symptoms can include stunting and swollen, fleshy, misshapen leaves. Infected plants should be removed and destroyed.

The growth abnormality on this painted daisy flower indicates a genetic abnormality in the plant.

Teucrium

Lamiaceae

Teucrium, or germander, is a genus of subshrubs known for their use as medicinal herbs, including some North American natives. Teucriums are also grown for their scented and attractive foliage in perennial gardens. One commonly grown species is *T. chamaedrys*, or wall germander. It will grow best in well-drained soils in full sun, but should be protected from drought. Neutral to alkaline soils are most desirable, and plant form may be best in soils of low fertility. Light shade is tolerated. *T. canadense* is an American native. Unlike other species of the genus, it does well in wet soils.

Three soil-dwelling fungi—*Rhizoctonia solani, Thielaviopsis basicola*, and *Sclerotium rolfsii*—can attack germander. Rhizoctonia root and stem rot causes roots to turn brown and makes stems wither at the base. Thielaviopsis attacks only the roots, but foliage may turn yellow or brown, and plants may be stunted or collapse. Strategies to reduce Rhizoctonia and Thielaviopsis root and stem rots include avoiding stressful growing conditions that can predispose plants to becoming infected, not reusing potting media, and avoiding replanting germander or other hosts of these fungi in landscape sites where these diseases have occurred. Several fungicides can be used as drench treatments, but another option is to remove and destroy infected plants. Southern blight, caused by *S. rolfsii*, causes the stem base to darken and soften; masses of white fungus may develop on the soil around the infection site, and numerous mustard-seed-size, white to brick red sclerotia may be embedded in the fungus near the stem. Affected stems typically wither and die. Southern blight risk can be reduced by prompt removal of infected plants. In the landscape, the top six inches of soil near infected

Rhizoctonia can start on germander as infected cuttings with slight brown spots.

plants can be removed and replaced with clean soil. Fungicide drenches are not recommended in the garden for this disease.

Downy mildew, caused by a primitive fungus-like organism, produces dark blotches on leaves, with fuzzy masses of grayish growth on the leaf undersides directly beneath the discolored areas. During prolonged rainy mild weather, downy mildew can blight large portions of the foliage. Cultural control begins with watering at the soil surface rather than from overhead, if possible, and spacing plants widely enough apart to ensure good air movement and rapid drying. Fungicides are available for downy mildew control, but generally need to be used as protectants rather than applying them after disease is noted.

Thalictrum

Ranunculaceae

Thalictrum, or meadow-rue, is a genus of medicinal herbs and ornamental plants including some North American natives. They are best suited to moist sites in rich soils, and should be provided with partial shade. These are mostly tall, graceful plants that should not be over-fertilized, and may need to be staked. Some are heat tolerant (*T. aquilegifolium*, columbine meadow-rue and *T. flavum*, yellow meadow-rue). For sites with full sun exposure and drier soil conditions, *T. flavum* subsp. *glaucum* is a good choice. Growth of *Thalictrum* is delayed in the spring.

Various kinds of leaf symptoms may appear on meadow-rue. Some of these are leaf spots caused by fungi in the genera *Ascochyta*, *Botrytis*, *Cercospora*, *Cercosporella*, *Cylindrosporium*, *Gloeosporium*, *Mycosphaerella*, *Pseudocercosporella* and *Septoria*. White smut disease caused by *Entyloma* spp. may be another cause for round leaf spots on *Thalictrum*. Leaf gall may be due to *Synchytrium*. Both powdery mildew (causing white colonies on the leaf surface) and downy mildew (causing pale leaf spots with a fuzz of sporulation on the undersurface of the leaf) are seen fairly commonly. Rust caused by various *Puccinia* and *Tranzschelia* species commonly occurs on many different *Thalictrum* species. Smut caused by *Urocystis sorosporioides* is also seen on this plant genus. Virus symptoms are seen infrequently. Ring spots may be caused by *Tobacco rattle virus* (TRV), which may be spread by certain species of nematodes (*Xiphinema* and *Longidorus*).

White ring spots form on meadow-rue infected with *Tobacco rattle virus* (TRV).

Powdery or downy mildew can cause yellow blotches to appear on upper surfaces of infected meadow-rue leaves.

Thermopsis

Fabaceae

Thermopsis, or false lupine, provides yellow flowers that resemble lupines in early spring. Some are native to North America. In hot, sunny climates, thermopsis may be planted in partial shade; otherwise it should be grown in full sun. Planting in light, well-drained fertile loam is recommended, but *Thermopsis* spp. will tolerate a range of soil conditions. Cut back the foliage a month after the blooming period has ended to avoid watching the leaf quality deteriorate. Thermopsis will not transplant readily, as it has a long taproot. This feature helps these plants to be drought tolerant.

Thermopsis is a low-maintenance perennial that is relatively free from diseases. Powdery mildew is the most common problem; it will cause white colonies or a light white coating across the leaf surface. Occasional leaf spots may be caused by fungi including *Ascochyta*, *Cercospora*, *Phoma*, *Phyllosticta*, *Ramularia* and *Stigmina* species. A rust caused by *Poliomopsis thermopsidis* forms its telial stage on *T. rhombifolia*. Reducing humidity around plants can help to suppress powdery mildew, and keeping water off leaves will help against rust and leaf spots. Powdery mildew and rust can be slowed by regularly removing diseased leaves, and fungicide sprays are an additional option. Mosaic and distortion of leaves may indicate infection with *Bean yellow mosaic virus* (BYMV). This virus may be spread by aphids to certain other plant species, as well as to other thermopsis plants, so plants with these symptoms should be removed from the garden.

Thymus

Lamiaceae

Thymus, or thyme, encompasses many species, including a culinary herb, *T. vulgaris*, as well as some ornamentals. Lemon thyme (*T.* ×*citriodorus*), woolly thyme (*T. pseudolanuginosus*), creeping thyme (*T. praecox*), and wild thyme (*T. serpyllum*) are all commonly grown. They are extremely attractive to bees and need full sun locations and excellent drainage. Rock garden areas or rock walls are ideal sites. Grow them in neutral to alkaline soils of moderate fertility.

more prone to rot during the summer in hot, humid southern climates than in cooler northern gardens. Frequent cutting back will help keep plants healthy and vigorous. Prune out areas of winter damage in the spring.

Minor problems with leaf spots on thyme may be due to a wide range of fungi including *Alternaria*, *Botrytis*, *Corynespora*, and *Mycosphaerella* species, as well as powdery mildew. Water molds (*Pythium*) may cause root rot, while the fungus *Rhizoctonia solani* is commonly responsible for crown rot. These diseases are all more common in the production nursery where overhead irrigation and close spacing of the plants make ideal conditions for disease initiation and spread. Avoid planting thyme in areas with poor soil drainage to discourage these root problems.

Crown rot can be caused by *Rhizoctonia solani* on thyme.

Powdery mildew on thyme appears as frosty white patches on upper leaf surfaces.

Tiarella

Saxifragaceae

Species of *Tiarella*, or foamflower, are woodland natives in North America, including Allegheny foamflower, *T. cordifolia,* from the East and three leaved foamflower, *T. trifoliata,* from the Pacific Coast. Western species are less tolerant of heat and humidity than *T. cordifolia* and its many varieties, which do well in the eastern United States. Tiarellas are easy to grow and, because they are stoloniferous, perform well as groundcovers in medium- to heavily-shaded sites. Full sun or excessive dryness will result in unhealthy-looking plants. Soils should ideally be humus-rich and retain moisture well, but excess moisture in winter is harmful to plants.

Several Puccinia rusts occur on foamflower. One of the most common is *Puccinia heucherae*, which also infects heuchera. Leaf spots are also sometimes seen, caused by fungi such as *Ramularia* or *Septoria* species, or by bacteria. The bacterial lesions can have a conspicuous yellow halo around the brown spots. A Synchytrium leaf gall is also known to occur on tiarella. Powdery mildew fungi may cause white colonies on the leaf surface. Proliferation of shoots at the root crown of a tiarella plant may be due to the bacterium *Rhodococcus fascians*.

Bacterial leaf spots on foamflower are dark brown to black and are wet looking.

Tradescantia

Commelinaceae

Tradescantia, or spiderwort, includes one major group of ornamental garden plants, the *Tradescantia virginiana* hybrids. *T. virginiana* is a North American native. Tradescantias should be grown in sunny, moist but well-drained areas of the garden. Flowering is reduced in partly shaded or overly wet soils, but light shade will be helpful for plant survival in hotter climates. Rejuvenate clumps by division every 2-3 years. Foliage may be cut back, leaving one foot of growth, immediately after flowering. Fertilize only lightly and plant in soils of lower fertility to keep vegetative growth relatively compact.

Diseases are rarely seen on this vigorous plant. Tradescantia is susceptible to leaf spots caused by the fungi *Alternaria*, *Cercospora*, *Colletotrichum*, *Cylindrosporium*, *Myrothecium*, *Phyllosticta*, *Pyricularia*, *Ramularia*, *Rhizoctonia* and *Septoria*, as well as to Botrytis blight. Rust symptoms on the foliage may be caused by *Uromyces commelinae*. These foliar problems will be kept to a minimum as long as plants are not given excessively frequent overhead watering. Galls on the roots may be due to root knot nematode (*Meloidogyne*), while root rots may be caused by *Fusarium*, *Pythium* or *Rhizoctonia*. As long as extremes of moisture supply are avoided, tradescantias should not be troubled by root rot.

Tricyrtis

Liliaceae

Tricyrtis, or toad lily, includes some curiously beautiful species of fall-blooming garden plants from the Orient: *T. formosana* and *T. hirta*. Hybrids are also available. Slightly acid soils rich in organic matter, partial shade and moisture are good for growing toad lilies. Summer mulches will help to keep their roots moist. Given time, tricyrtis will form large colonies; they are best left in place rather than disturbed.

A *Puccinia* rust, *Botrytis elliptica* leaf spot and an anthracnose due to *Colletotrichum* sp. have been reported for toad lily in Japan. In North America, the only reports of diseases of toad lily are *Lily virus X* (LVX), *Cucumber mosaic virus* (CMV) and *Tricyrtis virus Y* (in *T. formosana*). These viruses have been found occurring together in plants showing tiny pale ringspots and mosaic. Plants in the nursery or garden that display virus-like symptoms should be destroyed.

Mixed virus infection can result in white ringspots on toad lily.

Cucumber mosaic virus (CMV) was found in this mottled toad lily leaf.

Trillium

Liliaceae

Trilliums are woodland plants whose flowers have three sepals and three petals. They also occur in Asia, but the most ornamental species are native wildflowers in the United States, including *T. erectum*, stinking Benjamin, and *T. grandiflorum*, the great white trillium. Deeply or partly shaded conditions in moist, rich soil that is high in organic matter are best for trilliums. Soils that are acidic to neutral are generally preferred, although there are exceptions. The foliage may naturally die out before the end of summer. Mulching in the fall with leaves is recommended. Trilliums do well under trees with deep taproots, but they do not compete well for moisture with surface-rooted species.

Trilliums are not usually harmed by diseases. Under wet conditions in nurseries, however, they may be plagued with foliar nematodes (*Aphelenchoides*). Discard plants that show angular brown leaf spots that indicate foliar nematodes, and avoid introducing such plants to the garden. Reduce leaf wetness by altering overhead irrigation practices if foliar nematodes become a problem. Trillium leaves may occasionally be infected by a number of leaf spotting fungi: *Colletotrichum*, *Ciborinia*, *Gloeosporium*, *Heterosporium*, *Phyllosticta* and *Septoria* species. A Uromyces rust and a smut (*Urocystis trillii*) may affect the foliage as well. Manage foliar diseases by limiting overhead watering to morning applications and removing spotted leaves as soon as they are noticed.

Stem rot symptoms may be caused by *Sclerotium rolfsii*. This fungus produces white mycelium at the stem base of infected plants. Tiny, round, tan to dark brown sclerotia may be seen on the white fungus strands. Dig up and discard individual plants with stem rot, carefully removing a few inches of soil from the immediate area as well. Be careful not to spread the sclerotia. In nurseries, containers contaminated with *S. rolfsii* must be thoroughly cleaned before re-use to avoid spreading the fungus: it has a wide host range.

Trollius

Ranunculaceae

Trollius, or globeflower, features buttercup-like plants, some of them native to North America. Many varieties are sold as *T.* ×*cultorum* hybrids. Trollius should be grown in full sun or part shade, in very moist conditions. The soil should be both moisture-retentive and fertile. *Trollius* spp. may be grown along streams or in bog gardens—wherever they can be protected from heat or drought. The common globeflower, *T. europaeus*, is relatively tolerant of dry conditions. To encourage re-bloom, stems should be cut back and plants should be fertilized after blooming. The foliage will otherwise decline in attractiveness towards the middle of the growing season. Flowering is somewhat reduced in the first year after the shock of dividing and transplanting.

Trollius is not prone to diseases. Leaf spots on trollius may be caused by a few fungi, including *Ascochyta*, *Cylindrosporium montenegrium* and *Phyllosticta trollii*. Leaves and stems may be blistered by *Urocystis anemones*, a smut fungus. Avoid excessive overhead watering in order that these foliar diseases will not be fostered by cultural conditions in nursery or landscape. Trollius is also susceptible to *Cucumber mosaic virus* (CMV).

Tulipa

Liliaceae

Tulipa, or tulip, is a very familiar group of garden plants that offers immense diversity of flower form. There are thousands of hybrid varieties, as well as over a hundred species. Although perennials, the hybrid tulips generally fail to perform well for more than a few years, particularly in the South. High night temperatures take their toll on the bulbs, even if squirrels don't manage to find them. Treating them as annuals or very short-lived perennials is advisable. Full sun is desirable in the north, or at least afternoon sun. In the southern United States, morning sun or partial shade will give better results with tulips than full sun. They should be grown in fertile soils that are well-drained, and prefer neutral to somewhat acidic soil conditions.

Fire, or Botrytis blight, is the most common disease of tulips. The fungus *Botrytis tulipae* attacks all parts of tulip plants, and can cause increasing damage in a tulip bed year by year once it appears. The first evidence of the disease in the spring is often the appearance of scattered stunted shoots, also know as "fireheads," that emerge with blighted, twisted, tightly rolled leaves. The weakened shoots may collapse and die. In moist weather, a dense grayish brown fuzz of fungal growth develops on the diseased shoots. Small, dark green flecks soon appear on leaves, stems and flower buds of nearby tulips, then become oval to round and yellow to grayish brown with dark, water-soaked borders, blighting entire leaves and stems. Numerous black, pinhead-size sclerotia—the survival structures of *B. tulipae*—form in larger diseased spots. Buy or save only healthy-looking bulbs and select planting sites with good air and soil drainage. Avoid replanting tulips in the same location more often than every third year and discard all spotted, damaged, or moldy bulbs. Heavy mulches, high rates of nitrogen fertilizer, and overwatering should all be avoided. It is helpful to buy fungicide-treated bulbs, and fungicide sprays can be used to protect healthy plants.

Anthracnose, caused by the fungus *Gloeosporium thumenii*, can also cause brown spots on leaves. The same practices recommended against tulip fire should be effective against anthracnose.

Several other fungi attack tulip bulbs or roots. For example, *Sclerotium rolfsii* causes crown rot, resulting in masses of dense white fungal growth on diseased plant

Blighting of tulip leaves due to Botrytis infection.

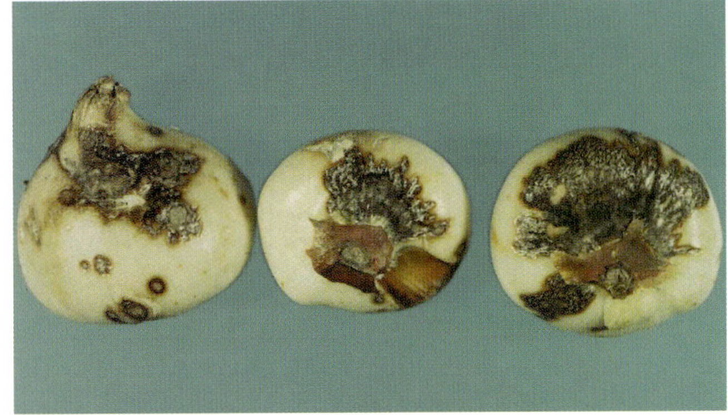

Tulip bulb lesions caused by *Botrytis tulipae*.

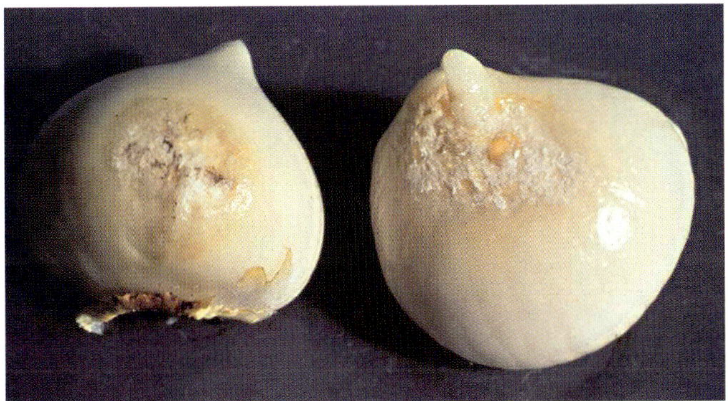

Fusarium oxysporum bulb rot on tulip.

A more advanced stage of Fusarium bulb rot on tulip.

Closeup of green *Trichoderma* sp. sporulation on infected tulip roots.

Tulip leaf withering on plants with *Trichoderma* and *Fusarium* spp. infecting their root systems.

Brown spots on tulip petals due to infection by *Botrytis tulipae*.

parts, speckled with numerous pinhead-size sclerotia that range from white to brick red. Basal rot, caused by a *Fusarium* species, produces large, dark brown spots on bulbs, accompanied by white or pink fungal growth. Blue mold, caused by *Penicillium*, results in bluish-or greenish-colored fungal growth on bulb surfaces and sometimes between scales. Black rot, also known as black slime, is caused by *Sclerotinia bulborum*; in roots and bulbs, it produces white masses of fungus and black, oblong sclerotia that may exceed ¼ inch in length. Additional bulb-rotting fungi include *Aspergillus* sp., *Rhizoctonia* spp., *Sclerotinia sativa*, and *Rhizopus stolonifer*. The fungus-like organisms *Phytophthora cactorum* and *Pythium* spp. can rot not only bulbs, but also roots and stems. To avoid fungal bulb rots, inspect all bulbs before buying and planting, discarding any that appear moldy, soft, or noticeably lighter than normal. Purchasing fungicide-treated bulbs provides an additional level of protection. If growing tulips in pots, make sure that the growing medium is steam-sterilized beforehand, and/or treat bulbs or media with fungicides. Remove and discard infected plants. Many bulb-rotting fungi have very durable survival structures, so replanting tulips in beds that have had serious bulb rot problems should be avoided for at least several years. Disinfest pots thoroughly before re-use.

Bulb rots can also be caused by bacteria. Soft rot, caused by the bacterium *Pectobacterium carotovorum*, prevents flowers from forming or blooming. Foliage may appear water-soaked and collapse. Infected bulbs may have a putrid odor and are soft and slimy. To reduce the risk of disease spread, remove infected foliage in the fall. Plant bulbs in well-drained soil, and water early in the day. Since bacteria are wound invaders, it is important to avoid wounding bulbs during planting and cultivation. Tools should be cleaned with a 10% dilution of household bleach, 70% alcohol, or

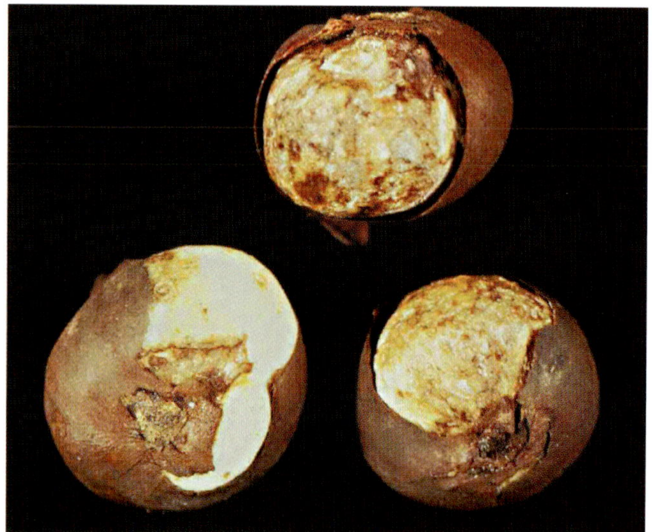

Spongy dead areas in tulip bulbs invaded by the nematode *Ditylenchus destructor*.

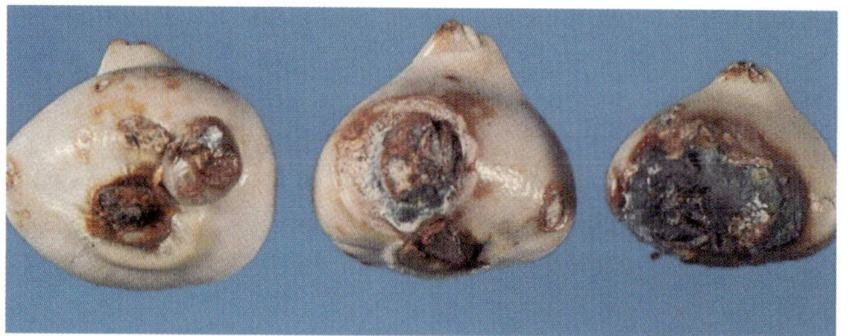

Blue sporulation of *Penicillium* sp. on areas of tulip bulb rot.

A tulip infected by a potyvirus shows pale, distorted foliage as well as flower break.

Distortion of tulip flower due to the nematode *Ditylenchus dipsaci*.

a commercially available sanitizer between uses. Controlling insects and mites can also reduce soft rot, since wounds they make can provide entries for bacterial infection.

Feeding by the stem and bulb nematode, *Ditylenchus dipsaci,* leaves bulbs with gray to brown patches that feel spongy. The bulbs are lighter in weight than normal, and the interior of the bulb is mealy in texture. To avoid these problems, purchase and plant bulbs that have been hot-water treated. Discard bulbs with suspicious symptoms.

Several viruses attack tulips, including tulip breaking (TBV), tobacco necrosis (TNV), tobacco rattle (TRV),

Fusarium bulb rot and associated stunting and scorch symptoms on above-ground portions of tulip.

White petal streaking (left) and color breaking (right) of flowers due to *Tulip breaking virus* (TBV).

tobacco mosaic (TMV), and cucumber mosaic (CMV). The best-known symptoms, termed breaking, appear as stripes or flecks on petals of normally solid-color tulips. Although breaking was once considered a highly desirable trait by tulip breeders and fanciers, it has since been recognized that virus-induced breaking gradually weakens plants. Flower shape may be distorted, and leaves may also become mottled by viruses. Infected plants should be removed and discarded to prevent aphids or other vectors from spreading the viruses to nearby tulips. Only virus-free bulbs should be planted. Some varieties of stripe-flowered tulips in the commercial trade resemble "broken" types but are not infected by a virus.

Distortion of tulip foliage by *Botrytis* sp. (left) and sclerotia of *Botrytis* sp. on tulip bulb (right).

Verbascum

Scrophulariaceae

Verbascum, or mullein, contains many species, including some that are quite ornamental. Garden species and hybrids tend to have woolly leaves and yellow, purple or white flower spikes. They are tolerant of poor (but well-drained) alkaline soils and do well in full sun. Alpine species should be protected against excessive soil moisture in winter. Verbascums are prone to spider mites.

Powdery mildew is one of the more common diseases on mullein. Fungal leaf spots caused by two species of *Cercospora*, as well as *Mycosphaerella*, *Phyllosticta*, *Pleospora*, and *Septoria*, have also been reported in North America. Stem cankers of mullein species due to various species of the fungus *Phoma* are seen fairly often. Root rot caused by *Phymatotrichopsis omnivora* has been noted in

Texas on a few species of verbascum, and Pythium root rot also occurs. There is also a report of a downy mildew on mullein in the northeastern United States. This disease is recognized by noting patches of fuzzy sporulation on the undersurfaces of leaves that show chlorotic spotting. Mulleins should be grown under cultural conditions that avoid long periods of continuous moisture on the foliage—to guard against fungal leaf spots and stem infections as well as powdery and downy mildew. Do not overcrowd plants, as there should be good air movement around them. Individual spotted leaves may be removed to reduce inoculum in the nursery or garden.

The mullein plant on the right is wilted and dying due to Pythium root rot.

A fungal leaf spot on mullein.

Verbena

Verbenaceae

Verbena, or vervain, is a genus that includes popular annual bedding plants. The perennial species and hybrids do well in full sun and moderately fertile, well-drained soils. *V. bonariensis*, *V. tenuisecta* and *V. rigida*, native to South America, have naturalized in the southeastern United States, and thus are obviously drought and heat tolerant. These plants are considered invasive in some areas, however. The rose verbena, *V. canadensis,* is often treated as an annual. Verbenas should be protected with a dry mulch in winter in areas of borderline hardiness.

The fungus *Botrytis cinerea* may cause a flower blight on verbena under rainy environmental conditions. If it is humid or plants are overcrowded, powdery mildew may be a significant problem, but susceptibility will vary greatly for different species and different varieties. Several Puccinia rusts have also been reported from verbenas. Choosing disease resistant varieties and species may be the best way to avoid having to treat verbenas with fungicides for protection against powdery mildew and rust. The roots or lower stems of verbena may be attacked by various fungi including *Rhizoctonia solani*, *Sclerotium bataticola*, *Thielaviopsis basicola*, and *Phymatotrichopsis omnivora*. *Pythium* and *Phytophthora* water molds can cause root rot under conditions of poor drainage. Strategies against root and stem rots include avoiding waterlogged soil conditions and wide swings in soil moisture availability, fungicide soil drenches, and removal of diseased plants.

Galls on the roots may be due to the root knot nematode, *Meloidogyne*. Verbena is susceptible to impatiens necrotic spot (INSV) and tomato spotted wilt (TSWV). Remove individual plants with virus or nematode problems from the garden or nursery, as these are not treatable.

Powdery mildew on verbena.

Fusarium stem rot on verbena.

Veronica

Scrophulariaceae

Veronica, or speedwell, includes some garden plants that grow well in sunny sites in light, moderately fertile, well-drained soil. Some tolerate afternoon shade. Gentian speedwell, *V. gentianoides*, tolerates moist soils. The woolly-leaved *V. incana* is not suited to hot humid climates with high rainfall, as the leaves are prone to diseases. Comb speedwell, *V. pectinata*, is more tolerant of dry sites than other species. Spiked speedwell, *Veronica spicata*, and its hybrids are common garden plants; they are easily grown if protected from wet soil conditions in winter. Deadhead regularly for repeated blooms.

Veronicas are especially vulnerable to the leafhopper-vectored phytoplasma disease aster yellows, and may develop virescent (green) flowers that show phyllody (leafy parts forming within the flower head). Another disease that changes the normal shape of the plant is shoot proliferation caused by infection by the bacterium *Rhodococcus fascians*. Discard plants with either of these symptoms. Veronicas are also prone to developing downy mildew, caused by *Peronospora grisea*, which typically causes them to exhibit systemic symptoms of chlorosis, stunting and distortion of the younger growth at the top of the plant in the spring. Look closely at the undersides of distorted leaves for the downy sporulation that gives this disease its name. Use overhead irrigation very cautiously on veronicas in the springtime to avoid fostering downy mildew. Watering thoroughly in late morning or at midday, rather than syringing with frequent light waterings, will suppress humidity around the plants. Avoid purchasing plants with leaf spots or stunted, distorted or pale young leaves that may indicate downy mildew.

Several fungal diseases are seen less commonly. Leaf galls caused by *Synchytrium* spp. and *Sorosphaera veronicae* are occasionally seen, as well as leaf spots caused by *Cercospora*, *Gloeosporium*, *Ramularia* and *Septoria*. White smut caused by an *Entyloma* species may also spot the foliage. Powdery mildew can cause white patches on the leaves, while several different Puccinia rusts may affect veronicas. Keeping overhead watering to a minimum will keep these in check. Root rot on veronicas may be due to attack from *Rhizoctonia solani*, *Fusarium* or *Phymatotrichopsis omnivora*, and *Fusarium* and *R. solani* may also cause a stem base canker, along with *Sclerotium rolfsii*. Stunted plants may be due to the action of root knot nematodes (*Meloidogyne*) feeding on the roots—look for the small galls or swellings on the root system to make a diagnosis. If root knot is found, be sure to replace the infested plants with species that are not hosts of *Meloidogyne*.

Veronica rust viewed from above resembles a fungal or bacterial leaf spot.

Bumpy pustules of rust on the undersurface of a veronica leaf.

Downy mildew on veronica may cause purpling of the upper leaf surface.

Abnormal, virescent flowers on veronica most likely caused by the aster yellows phytoplasma.

Powdery mildew has formed purplish patches on the upper surface of these veronica leaves.

Septoria leaf spot on veronica.

Viola

Violaceae

Viola, or violet, includes both the true violets (*Viola odorata*, for example) and the pansies (such as *Viola tricolor*). Hybrid pansies, (*V.* x *wittrockiana*) are used as annual bedding plants for the most part, even though they are perennial in some regions. True violets are best grown in shady, moist sites but will tolerate full sun if moisture is abundant. Horned violet, *Viola cornuta,* is one heat tolerant violet that does better than pansies in the south. Some of the *Viola* species are natives. Birdsfoot violet, *V. pedata*, for example, is native to eastern North America; it requires particularly good drainage as well as some shade.

If you notice pale yellow spots on leaves of violet (*Viola odorata*), turn them over to see if these leaves have a rust infection. The rust fungus *Puccinia violae* commonly forms a cluster of pustules, with white rims, out from which spill masses of yellow-orange spores. Other *Puccinia* species and a *Uromyces* sp. have also been reported causing rust diseases on *Viola* spp. The Uromyces rust has an Andropogon as an alternate host while the Puccinia rust has all of its life stages on its *Viola* hosts. Violets will continue to flower year after year in spite of the presence of rust. Gardeners and nurserymen should nonetheless avoid excessive overhead irrigation that would encourage rust infections and thus weaken the plants.

Another relatively common disease is identified by white, sometimes purple-rimmed, spots along with whitish, slightly swollen areas that form on petioles and leaf veins: this is scab, caused by the fungus *Sphaceloma violae*. This fungus affects birdsfoot violet (*Viola pedata*) as well as *V. odorata*, *V. cucullata* (marsh blue violet), *V. tricolor* (Johnny jump-up) and a number of other species.

The most common of the leaf spotting fungi is *Cercospora violae*, which is especially damaging on pansies. Tiny purplish spots on younger leaves continue to enlarge, maturing to dark purple to blackish spots 1/4" or more in diameter, with fringed margins. The larger spots are generally seen on the lowest, oldest leaves on the plant, which may also show a great deal of yellowing. Rainfall or irrigation landing on these leaf spots causes the spores to splash to new areas on the same or nearby plants' foliage, and the disease can spread quickly under wet weather conditions or during

Violet leaves infected with a rust fungus sometimes appear chlorotic from above.

On this violet leaf, pale green blisters caused by rust are just beginning to develop spore structures.

greenhouse production. Other leaf spots formed on violas on occasion may be due to fungi in the genera *Alternaria*, *Ascochyta*, *Botrytis*, *Centrospora*, *Ciborinia*, *Colletotrichum*, *Cryptostictis*, *Cylindrosporium*, *Heterosporium*, *Marssonina*, *Phyllosticta*, *Ramularia,* and *Septoria*.

Powdery mildew fungi can cause white patches on violet and pansy foliage, but this disease is common only in warm, humid growing conditions. In cool, moist climates, pansies are more often seen with purplish or yellowed foliage due to infection by a downy mildew, *Peronospora violae* or *Bremiella megasperma*; other violas are also susceptible to this disease. *Viola* spp. are also susceptible to aster yellows and beet curly top virus (BCTV). Proliferation of shoots due to infection by the bacterium *Rhodococcus fascians* also occurs on *Viola* spp.

Pansies and violas are especially susceptible to infection by the fungus *Thielaviopsis basicola*, which causes black root rot. This disease can cause stunting and death of plants. Soils with pH above 6.2 are very conducive to this disease; lowering pH is one technique for disease management. Improving drainage can also decrease the damage from Thielaviopsis root rot. Direct examination of the root system will show tiny brown resistant spores of the fungus that form within the roots. The other likely root and stem diseases are Pythium root rot, which softens and discolors the tips of the roots, and Rhizoctonia stem rot, which attacks roots or stem near the soil line, forming a brown stem canker that can either weaken or girdle the plant. In addition, *Sclerotium rolfsii* and *Phytophthora nicotianae* may also attack at the crown. *S. rolfsii* forms tiny mustard-seed sized structures called sclerotia that may be visible on the base of killed plants, along with white strands of the crown-rotting fungus, while *P. nicotianae* softens the crown tissue of pansies so much that wilting plants easily pull apart at the soil line when tugged upon. Fungicides may be used to protect plants against these diseases in the nursery. If any of these root and stem diseases are encountered during nursery production, the containers of diseased plants should not be re-used without a thorough disinfestation.

Cercospora leaf spot on pansy.

Powdery mildew on pansy.

White lesions of scab on violet leaves.

Scab on violet leaves, sepals and flower stalk.

Thielaviopsis signs (dark brown patches) on roots of pansy.

Thielaviopsis root rot causing stunt in a flat of pansies.

Waldsteinia

Rosaceae

Waldsteinia, or barren strawberry, is a small genus of plants including a few ornamental species. Barren strawberry, *W. fragarioides*, is native to the Eastern United States. It should be grown in full sun or partly shaded areas, where it provides evergreen edging. It does not do well in hot humid southern climates; *W. lobata* and *W. fragarioides* ssp. *doniana*, also natives, are preferable in the south. The Siberian barren strawberry, *W. ternata*, is a more attractive ground cover than *W. fragarioides* in cool climates.

Waldsteinia is generally untroubled by diseases. Occasionally leaf spots are caused by the fungi *Cercospora*, *Ramularia* and *Septoria;* there is also an anthracnose disease caused by *Gnomonia waldsteiniae* and a Synchytrium leaf gall. Waldsteinia is also susceptible to a Puccinia rust, as well as to a smut caused by *Urocystis waldsteiniae*.

Cercospora leaf spot on waldsteinia.

Yucca

Agavaceae

Yucca includes natives of the southwestern United States. These are very dependable plants, strong of form and of constitution. The most common species is *Y. filamentosa*, Adam's needle. Yuccas are grown in full sun, and require well-drained soils.

Leaf spots on yucca may be caused by fungi including *Cercospora*, *Coniothyrium*, *Cylindrosporium*, *Diplodia*, *Gloeosporium*, *Glomerella*, *Kellermania*, *Mycosphaerella*, *Phyllosticta*, *Stagonospora* and *Leptosphaeria*. Coniothyrium leaf spot is especially common, and will be most severe in rainy climates or if plants are regularly wetted by overhead irrigation. Warm wet conditions may also predispose yuccas to bacterial soft rot (*Pectobacterium* sp.). A Puccinia rust has been reported from Iowa and Nebraska.

Cultural conditions in nurseries may stress yuccas with excess moisture unless they are grown separately from plants that need more frequent irrigation. Planting in poorly drained sites or over-irrigating may predispose plants to infection by *Pectobacterium*, the cause of bacterial soft rot, or by *Sclerotium rolfsii*, which causes a fungal crown rot in the southern United States. *S. rolfsii* may be distinguished by the presence of small round sclerotia on or near the plant stem. Remove the infected plant and all evidence of the fungus if crown rot is seen. The plant is also susceptible to root knot nematode, *Meloidogyne*, which causes small galls to form like a chain of beads along the roots. If this disease is detected, replace with a type of plant that is not susceptible to root knot nematode.

Closeup of sporulation of *Coniothyrium* sp. on yucca.

Coniothyrium leaf spot is the most common disease on yucca.

Zantedeschia

Araceae

Zantedeschia, or calla lily, is a genus of South African natives. A few varieties are used in the garden, but the tubers are often dug and stored through the winter due to marginal hardiness in some zones. Callas should be thought of as "tropical perennials". *Zantedeschia aethiopica* is relatively trouble-free. Calla lilies grow best in partial shade, but can take full sun. They may also be planted in bog gardens, although heavy, waterlogged soils will predispose callas to soft rot. Air dry tubers in the sun before bringing them inside for winter storage. *Calla albo-maculata*, the spotted calla, has more cold tolerance than other species.

Calla lily is particularly vulnerable to infection by two bacteria that can cause a soft rot of tubers, stems and leaves, *Pectobacterium aroideae* and *P. carotovorum*. In greenhouse culture, careful sanitation practices are critical. Soft rot may originate in the tubers but sometimes appears as a root rot prior to shoot losses. The roots have a clear ("water-soaked") appearance when affected by soft rot bacteria.

All above-ground parts of calla lily are also susceptible to Botrytis blight. Occasional leaf spots are due to fungi such as *Cercospora, Coniothecium, Corynespora* and *Gloeosporium* species. The rust fungus *Uromyces ari-triphylli* causes a rust disease on calla as well as on *Arisaema* (Jack-in-the-pulpit). Reducing the length of leaf wetness intervals is helpful for minimizing leaf diseases on callas.

Under wet soil conditions or during container production, the tubers are vulnerable to infection by various species of *Phytophthora* and *Pythium*. Fungal root rots due to *Fusarium* and *Thielaviopsis* are also favored by wet soil, while both *Armillaria mellea* and *Rhizoctonia solani* can cause root rot of callas even in well-drained soils. Crown rot accompanied by small round sclerotia is due to the fungus *Sclerotium rolfsii*, which also does not require overwatering to be a problem. By far the most commonly encountered root disease in container production is caused by species of *Pythium*. Minimize irrigation and fertilization to produce callas without losses from Pythium root rot. Another type of root disease on callas is caused by nematodes feeding on roots: numerous small galls on the roots may be formed by root knot nematodes, *Meloidogyne*. There are no cures for root knot nematode infection; diseased plants should be disposed of, and the planting relocated.

Callas are also susceptible to two common thrips-borne viruses, *Impatiens necrotic spot virus* (INSV) and *Tomato spotted wilt virus* (TSWV). Various symptoms including leaf mottling and ring spots may be due to these problems. Occasionally *Dasheen mosaic virus* (DsMV) may also be encountered causing mosaic and leaf distortion in callas. Although samples should first be sent to a lab for identification, virus infected plants should be removed from production or garden areas, as no treatments are available. Identification of viruses will be helpful for learning what vectors and additional hosts might be relevant to the outbreak.

Mixed infection of *Alternaria* and *Pseudomonas* spp. on calla lily.

Stunting and flower break on a red-flowered calla with a mixed infection of *Tomato spotted wilt virus* (TSWV) and *Impatiens necrotic spot virus* (INSV).

Yellowing and wilting of callas with bacterial soft rot caused by *Pectobacterium* sp.

Botrytis blight on calla spathe.

Bacterial soft rot on calla tuber caused by a *Pectobacterium* species.

Common-Latin Name Directory

PLEASE NOTE: Plants are listed by their Latin names in this book, with the exceptions of ferns and ornamental grasses. See FERNS and GRASSES AND SEDGES, ORNAMENTAL in the text for common names.

COMMON NAME	LATIN NAME
Adam's needle	*Yucca*
African-lily	*Agapanthus*
Alumroot	*Heuchera*
Aster	*Aster*
Avens	*Geum*
Baby's breath	*Gypsophila*
Balloon flower	*Platycodon*
Baneberry	*Actaea* (previously *Cimicifuga*)
Barren strawberry	*Waldsteinia*
Barrenwort	*Epimedium*
Bear's breeches	*Acanthus*
Beardtongue	*Penstemon*
Beeblossom	*Gaura*
Begonia	*Begonia*
Bellflower	*Campanula*
Betony	*Stachys*
Bishop's weed	*Aegopodium*
Blackberry lily	*Belamcanda*
Blanket flower	*Gaillardia*
Blazing star	*Liatris*
Bleeding heart	*Dicentra*
Bluebells	*Mertensia*
Bugleweed	*Ajuga*
Buttercup	*Ranunculus*
Calla lily	*Zantedeschia*
Campion	*Lychnis*
Campion	*Silene*
Candytuft	*Iberis*
Canna lily	*Canna*
Catmint	*Nepeta*
Chinese lantern	*Physalis*
Chrysanthemum	*Chrysanthemum*
Cohosh	*Actaea* (previously *Cimicifuga*)
Columbine	*Aquilegia*
Coneflower	*Echinacea*
Coneflower	*Rudbeckia*
Confederate rose	*Hibiscus*

COMMON NAME	LATIN NAME
Cornflower	*Centaurea*
Corydalis	*Corydalis*
Cranesbill	*Geranium*
Crocosmia	*Crocosmia*
Daffodil	*Narcissus*
Dahlia	*Dahlia*
Daylily	*Hemerocallis*
Dead nettle	*Lamium*
Delphinium	*Delphinium*
Dill daisy	*Argyranthemum*
Elephant ear	*Caladium*
English daisy	*Bellis*
Euryops	*Euryops*
Evening Primrose	*Oenothera*
False indigo	*Baptisia*
False lupine	*Thermopsis*
False spirea	*Astilbe*
Foamflower	*Tiarella*
Forget-me-not	*Myosotis*
Foxglove	*Digitalis*
Foxtail lily	*Eremurus*
Gas plant	*Dictamnus*
Germander	*Teucrium*
Giant hyssop	*Agastache*
Gladiolus	*Gladiolus*
Globeflower	*Trollius*
Globethistle	*Echinops*
Goat's beard	*Aruncus*
Goldenrod	*Solidago*
Hellebore	*Helleborus*
Hen and chicks	*Sempervivum*
Hollyhock	*Alcea*
Hyacinth	*Hyacinthus*
Iris	*Iris*
Jack-in-the-pulpit	*Arisaema*
Jacob's ladder	*Polemonium*
Knapweed	*Centaurea*

COMMON NAME	LATIN NAME
Lady's mantle	*Alchemilla*
Lavender	*Lavandula*
Lavender cotton	*Santolina*
Leadwort	*Ceratostigma*
Leather flower	*Clematis*
Leopard plant	*Farfugium* (previously *Ligularia*)
Leopard's bane	*Doronicum*
Lewisia	*Lewisia*
Lily	*Lilium*
Lily-of-the-valley	*Convallaria*
Lily-turf	*Liriope*
Lobelia	*Lobelia*
Loosestrife	*Lythrum*
Loosestrife	*Lysimachia*
Lungwort	*Pulmonaria*
Lupine	*Lupinus*
Mallow	*Malva*
Marguerite daisy	*Argyranthemum*
Marsh marigold	*Caltha*
Masterwort	*Astrantia*
Mayapple	*Podophyllum*
Mayflower	*Maianthemum*
Meadow-rue	*Thalictrum*
Meadowsweet	*Filipendula*
Milkweed	*Asclepias*
Mint	*Mentha*
Monkshood	*Aconitum*
Mullein	*Verbascum*
New Zealand flax	*Phormium*
Obedient plant	*Physostegia*
Onion	*Allium*
Oregano	*Origanum*
Oswego tea	*Monarda*
Oxeye daisy	*Heliopsis*
Patrinia	*Patrinia*
Peony	*Paeonia*
Peruvian lily	*Astroemeria*
Phlox	*Phlox*
Pigsqueak	*Bergenia*
Pincushion flower	*Scabiosa*
Pink	*Dianthus*
Plantain lily	*Hosta*
Poppy	*Papaver*
Prickly pear	*Opuntia*
Primrose	*Primula*
Purple coneflower	*Echinacea*

COMMON NAME	LATIN NAME
Red valerian	*Centranthus*
Rockcress	*Arabis*
Rosemary	*Rosemarinus*
Rudbeckia	*Rudbeckia*
Rue	*Ruta*
Russian sage	*Perovskia*
Sage	*Salvia*
Saxifrage	*Saxifraga*
Sea holly	*Eryngium*
Sea-thrift	*Armeria*
Shasta daisy	*Leucanthemum*
Shrubby cinquefoil	*Dasiphora* (previously *Potentilla*)
Siberian bugloss	*Brunnera*
Skullcap	*Scutellaria*
Sneezeweed	*Helenium*
Solomon's seal	*Polygonatum*
Speedwell	*Veronica*
Spiderwort	*Tradescantia*
Spurge	*Euphorbia*
St. John's wort	*Hypericum*
Star flower	*Pentas*
Statice	*Limonium*
Stokes' aster	*Stokesia*
Stonecrop	*Sedum*
Strawberry	*Fragaria*
Sun rose	*Helianthemum*
Sundrops	*Oenothera*
Sunflower	*Helianthus*
Sweet woodruff	*Galium*
Tansy	*Tanacetum*
Thoroughwort	*Eupatorium*
Thyme	*Thymus*
Tickseed	*Coreopsis*
Toad lily	*Tricyrtis*
Torch lily	*Kniphofia*
Trillium	*Trillium*
Tulip	*Tulipa*
Turtle-head	*Chelone*
Vervain	*Verbena*
Violet	*Viola*
Wallflower	*Erysimum*
Wildginger	*Asarum*
Windflower	*Anemone*
Wormwood	*Artemisia*
Yarrow	*Achillea*

References

HERBACEOUS PERENNIALS

Armitage, A. M. 2008. Herbaceous Perennial Plants—A Treatise on Their Identification, Culture, and Garden Attributes, 3rd edition. Stipes Publishing, L.L.C., Champaign, IL.

Bennett, K.C. 2009. Pest Management Guide for the Production and Maintenance of Herbaceous Perennials. Cornell Cooperative Extension, Cornell University, Ithaca, NY. http://ipmguidelines.org/HerbaceousPerennials/default.asp

Bloom, A. 1974. Perennials for Your Garden, 3rd edition. Scribner, New York.

Bloom, A. 1977. Perennials in Island Beds: A Selection of the Best Hardy Perennials. Faber, London.

Brickell, C. and Zuk, J. 1977. The American Horticultural Society A-Z Encyclopedia of Garden Plants. DK Publishing, New York.

Brooklyn Botanic Garden. 2001. Native Perennials. Brooklyn Botanic Garden, Inc., Brooklyn, New York.

Burrell, C. C. 2002. The Sunny Border: Sun-Loving Perennials for Season-Long Color. Brooklyn Botanic Garden, Inc., Brooklyn, New York.

Burrell, C. C. and Burrell, C. 2002. The Shady Border: Shade-Loving Perennials for Season-Long Color. Brooklyn Botanic Garden, Inc., Brooklyn, New York.

Clausen, R. R. and Ekstrom, N. H. 1989. Perennials for American Gardens: the Definitive A-Z Reference Guide to Over 3,000 Species, Cultivars and Hybrids for Gardeners Across the Country. Random House, Westminster, MD.

Cohen, S. and Ondra, N. 2005. The Perennial Gardener's Design Primer—the Essential Guide to Creating Simply Sensational Gardens. Storey Publishing, North Adams, MA.

Ellis, B. 2000. Taylor's Guide to Perennials: More Than 600 Flowering and Foliage Plants, Including Ferns and Ornamental Grasses. Houghton Mifflin Company. New York.

Harper, P., McGourty, M. A. 1985. Perennials—How to Select, Grow and Enjoy. HP Books, Inc., New York.

Hawthorne, L. 2000. Perennials. DK Publishing, Inc., London.

Lloyd, C. 2002. Christopher Lloyd's Garden Flowers. Timber Press, Portland, OR.

Nau, J. 1996. The Ball Perennial Manual: Propagation and Production. Ball Publishing, Batavia, IL.

Pilon, P. 2006. Perennial solutions: a grower's guide to perennial production. Ball Publishing, Batavia, IL.

Still, S. 1993. Manual of Herbaceous Ornamental Plants, 4th ed. Stipes Publishing Company, Champaign, IL.

USDA, NRCS. The PLANTS Database. National Plant Data Center, Baton Rouge, LA. http://plants.usda.gov

FUNGAL LEAF SPOTS

Chase, A. R. 1997. Foliage Plant Diseases: Diagnosis and Control. APS Press, St. Paul, MN.

Chase, A. R., Daughtrey, M. L., and Simone, G. W. 1995. Diseases of Annuals and Perennials. Ball Publishing, Batavia, IL.

Daughtrey, M. L., Wick, R. L., and Peterson, J. L. 1995. Compendium of Flowering Potted Plant Diseases. APS Press, St. Paul, MN.

Farr, D. F., Bills, G. F., Chamuris, G. P., and Rossman, A. Y. 1989. Fungi on Plants and Plant Products in the United States. APS Press, St. Paul, MN.

Farr, D. F., and Rossman, A.Y. Fungal Databases, Systematic Mycology and Microbiology Laboratory, ARS, USDA. http://nt.ars-grin.gov/fungaldatabases/

Forsberg, J. L. 1975. Diseases of Ornamental Plants. Special Publication No. 3 Revised. University of Illinois Press, Urbana-Champaign, IL.

Gill, S., Cloyd, R. A., Baker, J., Clement, D. L., and Dutky, E. 2006. Pests and Diseases of Herbaceous Perennials—The Biological Approach. 2nd edition. Ball Publishing, Batavia, IL.

Horst, R. K. 2001. Westcott's Plant Disease Handbook, 6th ed. Kluwer Academic Publishers, Boston.

Powell, C. C., and Lindquist, R. K. 1997. Ball Pest and Disease Manual. 2nd edition. Ball Publishing, Batavia, IL.

Schumann, G. L. and D'Arcy, C. J. 2006. Essential Plant Pathology. APS Press, St. Paul, MN.

GRAY MOLD

Hausbeck, M. K., and Pennypacker, S. P. 1991. Influence of grower activity and disease incidence on concentrations of airborne conidia of *Botrytis cinerea* among geranium stock plants. Plant Disease 75:798-803.

Hausbeck, M. K., and Moorman, G. W. 1996. Managing *Botrytis* in greenhouse-grown flower crops. Plant Disease 80:1212-1219.

Jarvis, W. R. 1977. *Botryotinia* and *Botrytis* species: taxonomy, physiology, and pathogenicity. Vol. 15, Canada Department of Agriculture Monographs. Ottawa: Canada Department of Agriculture. 195 pp.

Sirjusingh, C., and Sutton, J. C. 1996. Effects of wetness duration and temperature on infection of geranium by *Botrytis cinerea*. Plant Disease 80:160-165.

POWDERY MILDEW

Braun, U., Cook, R. T. A., Inman, A. J., and Shin, H. -D. 2002. The taxonomy of powdery mildew fungi. In The Powdery Mildews: A Comprehensive Treatise, edited by R. Bélanger, W. R. Bushnell, A. J. Dik and T. L. W. Carver. APS Press, St. Paul, MN.

Jarvis, W. R., Gubler, W. D., and Grove, G. G. 2002. Epidemiology of powdery mildews in agricultural pathosystems. In The Powdery Mildews: A Comprehensive Treatise, edited by R. Bélanger, W. R. Bushnell, A. J. Dik and T. L. W. Carver. APS Press, St. Paul, MN.

Schumann, G. L. and D'Arcy, C. J. 2006. Essential Plant Pathology. APS Press, St. Paul, MN.

Trigiano, R. N., Windham, M. T. and Windham, A. S. 2003. Plant Pathology: Concepts and Laboratory Exercises. CRC Press, Boca Raton.

ROOT ROTS CAUSED BY FUNGI

Agrios, G. N. 2005. Plant Pathology. 5th ed. Academic Press, New York.

Baker, R., and Martinson, C. A., eds. 1970. Epidemiology of diseases caused by *Rhizoctonia solani*. *Rhizoctonia solani*, Biology and Pathology, J. R. Parmeter, ed. Univ. of California Press, Berkeley.

Chang, K. F., Barr, D. J. S., Hwang, S. F., and Mirza, M. 1994. Effect of interactions between *Fusarium*, *Rhizoctonia*, and *Pythium* on root and rhizome rot of *Alstroemeria*. Zeitschrift Fur Pflanzenkrankheiten und Pflanzenschutz—Journal of Plant Diseases and Protection 101:460-466.

Chase, A. R., Daughtrey, M. L., and Simone, G. W. 1995. Diseases of annuals and perennials. Ball Publishing, Batavia, IL.

Daughtrey, M. L., and Benson, D. M. 2005. Principles of plant health management for ornamental plants. Annual Review of Phytopathology 43:141-169.

Daughtrey, M. L., Wick, R. L., and Peterson, J. L. 1995. Compendium of Flowering Potted Plant Diseases. APS Press, St. Paul, MN.

Farr, D. F., Bills, G. F., Chamuris, G. P., and Rossman, A. Y. 1989. Fungi on Plants and Plant Products in the United States. APS Press, St. Paul, MN.

Farr, D. F., and Rossman, A.Y. Fungal Databases, Systematic Mycology and Microbiology Laboratory, ARS, USDA. http://nt.ars-grin.gov/fungaldatabases/

Forsberg, J. L. 1975. Diseases of Ornamental Plants. Special Publication No. 3 Revised. University of Illinois Press, Urbana-Champaign, IL.

Linderman, R. G. 1985. Root rots. In Diseases of Floral Crops, D. L. Strider, ed. Praeger Scientific, New York, NY.

Schumann, G. L. and D'Arcy, C. J. 2006. Essential Plant Pathology. APS Press, St. Paul, MN.

RUSTS AND SMUTS

Chase, A. 2002. Controlling rust diseases. Chase News 1:1-6.

Farr, D. F., Bills, G. F., Chamuris, G. P., and Rossman, A. Y. 1989. Fungi on Plants and Plant Products in the United States. APS Press, St. Paul, MN.

Farr, D. F., and Rossman, A.Y. Fungal Databases, Systematic Mycology and Microbiology Laboratory, ARS, USDA. http://nt.ars-grin.gov/fungaldatabases/

Forsberg, J. L. 1975. Diseases of Ornamental Plants. Special Publication No. 3 Revised. University of Illinois Press, Urbana-Champaign, IL.

Jones, R. K. and Benson, D. M., eds. 2001. Diseases of Woody Ornamentals and Trees in Nurseries. 2001. APS Press, St. Paul, MN.

Schumann, G. L. and D'Arcy, C. J. 2006. Essential Plant Pathology. APS Press, St. Paul, MN.

VASCULAR WILTS CAUSED BY FUNGI

Daughtrey, M. L., Wick, R. L., and Peterson, J. L. 1995. Compendium of Flowering Potted Plant Diseases. APS Press, St. Paul, MN.

Forsberg, J. L. 1975. Diseases of Ornamental Plants. Special Publication No. 3 Revised. University of Illinois Press, Urbana-Champaign, IL.

Nameth, S. T., Daughtrey, M. L., Moorman, G. W., and Sulzinski, M. A. 1999. Bacterial blight of geranium: A history of diagnostic challenges. Plant Disease 83:204-212.

Nelson, P. E., Tammen, J., and Baker, R. 1960. Control of vascular wilt diseases of carnation by culture-indexing. Phytopathology 50:356-360.

Dimock, A. W. 1943. A method of establishing *Verticillium*-free clones of perennial plants. Phytopathology 33:3.

Katan, J., Greenberger, A. , Alon, H. , and Grinstein, A. 1976. Solar heating by polyethylene mulching for the control of diseases caused by soil-borne pathogens. Phytopathology 66:683-688.

Schumann, G. L. and D'Arcy, C. J. 2006. Essential Plant Pathology. APS Press, St. Paul, MN.

DOWNY MILDEWS

Agrios, G. N. 2005. Plant Pathology. 5th ed. Academic Press, New York.

Farr, D. F., Bills, G. F., Chamuris, G. P., and Rossman, A. Y. 1989. Fungi on Plants and Plant Products in the United States. APS Press, St. Paul, MN.

Farr, D. F., and Rossman, A.Y. Fungal Databases, Systematic Mycology and Microbiology Laboratory, ARS, USDA. http://nt.ars-grin.gov/fungaldatabases/

Schumann, G. L. and D'Arcy, C. J. 2006. Essential Plant Pathology. APS Press, St. Paul, MN.

Yarwood, C. E. 1947. Snapdragon downy mildew. Hilgardia 17:241-250.

ROOT AND STEM ROTS CAUSED BY FUNGUS-LIKE ORGANISMS (*PYTHIUM* AND *PHYTOPHTHORA*)

Farr, D. F., Bills, G. F., Chamuris, G. P., and Rossman, A. Y. 1989. Fungi on Plants and Plant Products in the United States. APS Press, St. Paul, MN.

Farr, D. F., and Rossman, A.Y. Fungal Databases, Systematic Mycology and Microbiology Laboratory, ARS, USDA. http://nt.ars-grin.gov/fungaldatabases/

Gardiner, R. B., Jarvis, W. R., and Shipp, J. L. 1990. Ingestion of *Pythium* spp. by larvae of the fungus gnat *Bradysia impatiens* (Diptera, Sciaridae). Annals of Applied Biology 116:205-212.

Goldberg, N. P., and Stanghellini, M. E. 1990. Ingestion-egestion and aerial transmission of *Pythium aphanidermatum* by shore flies (Ephydrinae: *Scatella stagnalis*). Phytopathology 80:1244-1246.

Hoitink, H. A. J., and Boehm, M. J. 1999. Biocontrol within the context of soil microbial communities: A substrate-dependent phenomenon. Annual Review of Phytopathology 37:427-446.

Hong, C. X., and Moorman, G. W. 2005. Plant pathogens in irrigation water: Challenges and opportunities. Critical Reviews in Plant Science 24:1889-1908.

Moorman, G. W., and Kim, S. H. 2004. Species of *Pythium* from greenhouses in Pennsylvania exhibit resistance to propamocarb and mefenoxam. Plant Disease 88:630-632.

Moorman, G. W., Kang, S., Geiser, D. M., and Kim, S. H. 2002. Identification and characterization of *Pythium* species associated with greenhouse floral crops in Pennsylvania. Plant Disease 86:1227-1231.

Oyler, E., and Bewley, W. F. 1937. A disease of cultivated heaths caused by *Phytophthora cinnamomi* Rands. Annals of Applied Biology. 24:1-16.

Van West, P., Morris, B. M., Reid, B., Appiah, A. A., Osborne, M. C., Campbell, T. A., Shepherd, S. J., and Gow, N. A. R. 2002. Oomycete plant pathogens use electric fields to target roots. Molecular Plant-Microbe Interactions 15 (8):790-798.

BACTERIAL DISEASES

Agrios, G. N. 2005. Plant Pathology. 5th ed. Academic Press, New York.

Cooksey, D. A. 1990. Genetics of bactericide resistance in plant pathogenic bacteria. Annual Review of Phytopathology 28:201-219.

Dimock, A. W. 1962. Obtaining pathogen-free stock by cultured cutting technique. Phytopathology 52:1239-1241.

Drotleff, L. 2005. Certifiable perennials. Greenhouse Grower 25 (11):8, 10.

Forsberg, J. L. 1975. Diseases of Ornamental Plants. Special Publication No. 3 Revised. University of Illinois Press, Urbana-Champaign, IL.

Haygood, R. A., and Strider, D. L. 1985. Bacterial diseases. In Diseases of Floral Crops, D. L. Strider, ed. Praeger Scientific, New York.

PHYTOPLASMAS

Agrios, G. N. 2005. Plant Pathology. 5th ed. Academic Press, New York.

Bai et al. 2006. Living with genome instability: the adaptation of phytoplasmas to diverse environments of their insect and plant hosts. Journal of Bacteriology 188:3682-3696.

Beanland, L., Hoy, C. W., Miller, S. A., and Nault, L. R. 2000. Influence of aster yellows phytoplasma on the fitness of aster leafhopper (Homoptera: Cicadellidae). Annals of Entomological Society of America 93:271-276.

Lee, I.-M., Martini, M., Bottner, K. D., Dane, R. A., Black, M. C., and Troxclair, N. 2003. Ecological implications from a molecular analysis of phytoplasmas involved in an aster yellows epidemic in various crops in Texas. Phytopathology 93:1368-

VIRUS DISEASES

Agrios, G. N. 2005. Plant Pathology. 5th ed. Academic Press, New York.

Brunt, A.A., Crabtree, K., Dallwitz, M.J., Gibbs, A.J., Watson, L. and Zurcher, E.J. (eds.) (1996 onwards). Plant Viruses Online: Descriptions and Lists from the VIDE Database. http://www.image.fs.uidaho.edu/vide/refs.htm

Chase, A. R., Daughtrey, M. L., and Simone, G. W. 1995. Diseases of annuals and perennials. Ball Publishing, Batavia, IL.

Copes, W. E., and Hendrix, F. F. 1996. Chemical disinfestation of greenhouse growing surface materials contaminated with *Thielaviopsis basicola*. Plant Disease 80:885-886.

De Boer, S. H., Andrews, J. H., and Tommerup, I. C., eds. 1996. Pathogen Indexing Technologies, Advances in Botanical Research Vol. 23, J. A. Callow, ed. Academic Press, New York.

Drotleff, L. 2005. Certifiable perennials. Greenhouse Grower 25 (11):8, 10.

Grand, L. F. 1977. North Carolina plant disease index. Tech. Bul. No. 240. Raleigh, NC. North Carolina Agricultural Experiment Station.

Horst, R. K. 1987. Diagnostic methods for viruses of ornamental crops. In Diagnostic methods for viruses of ornamental crops. J. Hammond and R. H. Lawson, eds. ISHS. International Working Group on Virus Diseases of Ornamental Plants, Beltsville, MD.

ICTVdB website, the universal virus database of the International Committee on the Taxonomy of Viruses (ICTV): http://www.ncbi.nlm.nih.gov/ICTVdb/index.htm

Munnecke, D. E. 1956. Development and production of pathogen-free geranium propagative material. Plant Disease Reporter Supplement 238:93-95.

Paludan, N. 1985. Spread of viruses by recirculated nutrient solutions in soilless cultures. Tidsskr. Planteavl. 89:467-474.

Paludan, Von N. 1992. Testing the efficiency of disinfectants for control of virus: Results achieved in disinfection experiments and proposal for proofing guidelines. Nachrichtenbl. Deut. Pflanzenschutzd. 44:73-79.

Schumann, G. L. and D'Arcy, C. J. 2006. Essential Plant Pathology. APS Press, St. Paul, MN.

NEMATODES

Agrios, G. N. 2005. Plant Pathology. 5th ed. Academic Press, New York.

Schumann, G. L. and D'Arcy, C. J. 2006. Essential Plant Pathology. APS Press, St. Paul, MN.

Trigiano, R. N., Windham, M. T. and Windham, A. S. 2003. Plant Pathology: Concepts and Laboratory Exercises. CRC Press, Boca Raton.

Photo Credits

All images were supplied by Margery Daughtrey, Ann Chase, Gary Moorman or Daren Mueller, except for the following:

Allan Armitage, University of Georgia: p. 30, *Actaea*; p. 216, santolina in flower; p. 219, *Scutellaria*; **Larry Barnes, Texas A & M:** p. 160, *Macrophomina* (2); **Janna Beckerman, Purdue University:** p. 5, *Ascochyta*; p. 21, aster yellows; p. 34, AMV (older leaf); p. 36, rust; p. 48, rust (left); p. 77, *Ascochyta* (2); p. 78, ToRSV; p. 80, anthracnose; p. 81, aster yellows (multiple small flower heads); p. 91, TSWV; p.95, TRV (left); p. 106, *Phytophthora*; p. 111, powdery mildew; p. 116, *Pythium*; p. 118, *Xanthomonas* (lower image); p. 127, ergot; p. 139, *Rhizoctonia*; p. 140, p. 171, aster yellows; p. 172, anthracnose (closeup at left); p. 179, TSWV and rust; p. 186, INSV; p. 187, *Botrytis*; p. 188, *Cladosporium* (lower image, stem); p. 203, rust (2); p. 221, *Fusarium* and *Pythium*; p. 224, aster yellows; p. 229, TRV; p. 231, *Rhizoctonia*, bacterial leaf spot; p. 237, tulip breaking (image at left); p. 241, *Septoria*; **Central Science Laboratory, UK, Crown Copyright, Bugwood.org:** p. 74, *Phoma*; **William M. Ciesla, Forest Health Management International, Bugwood.org:** p. 151, *Iris cristata*; p. 176, *Maianthemum* (image on right); p. 223, *Silene*; **Andrea Clark, Farmingdale State:** p. 68, *Canna* (2); p. 81, powdery mildew (2); **Clemson University – USDA Cooperative Extension Slide Series, Bugwood.org:** p. 235, *Botrytis tulipae* on petals; **Don Cooksey, University of California, Riverside:** p. 128, *Pantoea*; **Tom Creswell, NC State University:** p. 6, *Alternaria*; p. 7, *Botrytis*; p. 38, *Sclerotium*; p. 76, TSWV; p. 90, *Sclerotium* (field) and *Botrytis*; p. 100, *Sclerotium*; p. 108, *Alternaria*; p. 110, *Rhizoctonia*; p. 152, *Pectobacterium*; p. 153, *Heterosporium* (*Mycosphaerella* stage, at left); p. 154, *Sclerotium* (2); p. 171, *Sclerotinia*; p. 181, *Stagonospora* (aboveground symptoms); p. 182, *Botrytis*; p. 192, CMV; p. 205, downy mildew; p. 226, TSWV; p. 233, CMV; p. 235, *Trichoderma* on roots, *Trichoderma* and *Fusarium*, Botrytis blight (leaves); **Toon Derks, WUR, Applied Plant Research - Bulbs, Trees and Fruit Section, Lisse, The Netherlands:** p. 155, iris yellow spot on iris; **Sharon Douglas, Connecticut Agricultural Experiment Station:** p. 42, bacterial leaf spot and foliar nematode; p. 52, INSV; p. 140, bacterial leaf spot; p. 196, *Septoria*; **Brooke Edmunds, Colorado State University Extension:** p. 206, *Sclerotium*; **Jon Eisenback, Virginia Tech:** p. 24, nematode mouthparts, nematode feeding, root knot injury; **Patti Ek, APS PRESS:** p. 151, *Iris versicolor*; **Florida Dept. of Agriculture and Consumer Services, Division of Plant Industry Archive, Bugwood.org;** p. 75, crown gall; **Dan Gilrein, Cornell University Cooperative Extension of Suffolk County, NY:** p. 99, insect injury; **Greg Grahek, APS PRESS:** p. 104, *Eupatorium*; **Mary Francis Heimann, University of Wisconsin:** p. 74, *Alternaria*; p. 182, *Botryotinia*; **Bill Hilbert, Kansas Department of Agriculture:** p. 142, foliar nematode; **Gordon Holcomb, Louisiana State University:** p. 99, CMV; p. 109, foliar nematode (upper right image), p. 169, anthracnose, *Phytophthora* (in ground, in flower bed); **Chuan Hong, Virginia Tech:** p. 49, *Phytophthora*; p. 221, *Pectobacterium* (image on right); **Steve Jeffers, Clemson University:** p. 45, copper fungicide injury; p. 108, *Colletotrichum*; p. 138, *Phytophthora*; p. 143, *Phytophthora*; p. 150, *Pythium*; p. 153, rust (early and mature); p. 158, *Myrothecium*; p. 169, *Phytophthora* in container; p. 196, black root rot; **Tony Keinath, Clemson University:** p. 82, *Phyllosticta*; **Caroline Kiang, Cornell University Cooperative Extension of Suffolk County, NY:** p. 149, *Iberis* (on right); p. 167, *Limonium*; p. 174, *Lychnis*; **Steve Koike, University of California-ANR:** p. 32, *Botrytis*, herbicide injury; p. 54, powdery mildew; p. 78, rust; **Jadwiga Komorowska-Jedrys, Cornell University:** p. 17 and p. 63, *Phytophthora*; **Jim LaMondia, Connecticut Agricultural Experiment Station, Windsor:** p. 30, *Meloidogyne*; p. 57 and p. 63, root knot nematode; **Ben Lockhart, University of Minnesota:** p. 52, unknown virus; p. 140, TRV; p. 145, TRSV; p. 158, caulimovirus; **Bent Loschenkohl, Odense, Denmark:** p. 66, *Phoma*; **Nancy Lowenstein, Auburn University, Bugwood.org:** p. 151, *Iris pseudacorus*; **Barbara Ludlow, Bridgehampton, NY:** p. 212, *Fusarium*; **Bob McGovern, University of Florida:** p. 64, *Pectobacterium*, *Pythium*; p. 65, *Fusarium*; **Barry Menser, Michigan Department of Agriculture:** p. 81, aster yellows (middle and lowest image); p. 98, aster yellows (image on far right); p. 138, TRSV; p. 145, HVX (middle right image); p. 182, potyvirus; p. 211, TSV; p. 236, potyvirus; **Donna Moramarco, Locust Valley, NY:** p. 101, TRV; **Bob Mulrooney, University of Delaware:** p. 22, CMV; p. 25, foliar nematode; p. 43, *Septoria*; p. 51, *Botrytis*, anthracnose; p. 52, *Thielaviopsis*; p. 53, foliar nematode (on right); p. 72, foliar nematode; p. 80, *Ascochyta*; p. 109, foliar nematode (lower left image); p. 110, *Taphrina* (2); p. 125, *Stagonospora* (3); p. 136, daylily leaf streak (left image); p. 141, *Fusarium*; p. 164, powdery mildew; p. 171, CMV(2); **Yoshitaka Ono, Ibaraki University, Mito, Japan. Fig. 1-C, p. 241 in Does *Puccinia hemerocallidis* regularly host-alternate between *Hemerocallis* and *Patrinia* plants in Japan? J. Gen. Plant Pathol. Vol 69(4):240-243. Copyright The Phytopathological Society of Japan and Springer-Verlag Tokyo 2003,**

with kind permission Springer Science and Business Media: p. 137 and p. 190, daylily rust on patrinia; **Theo Overdevest, Plant Protection Service of The Netherlands:** p. 14, *Verticillium*; p. 19, *Rhodococcus*; p. 29, *Aphelenchoides*, *Verticillium* and *Meloidogyne*; p. 38, *Sclerotinia cepivorum*; p. 39, *Fusarium*, *Ditylenchus* and *Aphelenchoides*; p. 60, soft rot; p. 68, *Botrytis* (2); p. 79, *Aphelenchoides*, *Pratylenchus* and *Puccinia*; p. 84, *Dickeya*, *Agrobacterium*/*Rhodococcus*; p. 85, *Rhodococcus*, *Ditylenchus*, *Meloidogyne* and *Sclerotinia*; p. 86, *Ralstonia*, TSWV (2); p. 102, *Pythium* and *Fusarium*, *Sclerotium*; p. 120, *Burkholderia*, *Rhodococcus*; p. 121, *Botrytis*, *Septoria*, *Stromatinia*, *Ditylenchus*, *Penicillium* and *Fusarium*; p. 122, CMV (2); p. 146, *Fusarium*, *Xanthomonas*, *Pectobacterium*, *Aspergillus*; p. 147, stem and bulb nematode; p. 154, *Drechslera*, *Sclerotium*, *Fusarium*; p. 163, *Botrytis*/*Sclerotinia*, *Verticillium* (2); p. 164, stem and bulb nematode; p. 165, *Fusarium*/*Cylindrocarpon*, *Sclerotium*; p. 181, *Ditylenchus*, *Botrytis*/*Stagonospora*; p. 235, Fusarium bulb rot (2), *Botrytis tulipae* (bulbs); p. 236, *Ditylenchus* (2), *Penicillium*; **Hanu Pappu, Washington State University:** p. 86, INSV; p. 87, DMV (2); p. 155: iris yellow spot on onion (2); **Leanne Pundt, University of Connecticut:** p. 2, lavender; p. 39, *Alstroemeria*; p. 53, *Asclepias*; p. 77, *Clematis*; p. 97, *Echinacea*; p. 140, rust; p. 145, INSV; p. 159, *Lavandula*; p. 163, *Liatris*; p. 190, *Patrinia*; p. 191, *Penstemon*; p. 192, INSV and rust (undersurface of leaf); p. 200, *Platycodon*, p. 241, rust (leaf undersurface); p. 245, *Yucca*; **Melodie Putnam, Oregon State University:** p. 27, *Rhodococcus* (2); p. 150, *Rhodococcus*; p. 216, winter injury; **Karen Rane, University of Maryland:** p. 76, *Pseudomonas*; p. 109, foliar nematode (lower right image); p. 188, *Cladosporium* (leaf); **Naidu Rayapati, Washington State University:** p. 82, lettuce mosaic; p. 179, INSV (2); **Gail Ruhl, Purdue University:** p. 44, CMV; **Guido Schnabel, Clemson University:** p. 138, *Armillaria* (2); **Cheryl Smith, University of New Hampshire:** p. 20, phytoplasma; p. 43, bacterial leaf spot; p. 44, leaf miner; p. 89, bacterial leaf spot; p. 92, *Botrytis*; p. 98, aster yellows (left and middle image); p. 135, downy mildew; p. 136, daylily leaf streak (image on right); p. 166, *Botrytis* on foliage; p. 172, anthracnose on stem; p. 180, powdery mildew; p. 187, frost damage; p. 203, powdery mildew (image on right); p. 209, powdery mildew; **Mike Tiffany, Agdia, Inc, Elkhart, IN:** p. 145, ToRSV, ArMV; **Maria Tobiasz, Cornell University:** p. 40, *Sclerotium* (2); p. 42, foliar nematode (early); p. 115, *Thielaviopsis* and *Rhizoctonia*; p. 150, downy mildew; p. 165, *Lilium*; p. 195, *Fusarium*, downy mildew; p. 234, *Tulipa*; p. 240, *Fusarium*; **Rodrigo Valverde, Louisiana State University:** p. 110, virus; **Colleen Warfield, Univ. of California-ANR:** p. 109, foliar nematode (upper left image); p. 133, foliar nematode; **Rob Wick, University of Massachusetts:** p. 53, foliar nematode; p. 158, *Phytophthora citrophthora*; p. 196, stem and bulb nematode; **Jean Williams-Woodward, University of Georgia:** p. 19, soft rot; p. 144, *Cercospora*, *Pectobacterium*; **Jim Willmott, Griffin Greenhouse and Nursery Supplies, Inc.:** p. 144, *Colletotrichum*; p. 150; *Leptosphaeria* (2); **Alan Windham, University of Tennessee:** p. 22, HVX; p. 100, unidentified virus; p. 145, HVX (lower right); p. 201, *Sclerotium*; **Rick Yates, Griffin Greenhouse and Nursery Supplies, Inc.:** p. 205, *Botrytis*.

Glossary

A

Actinomycete—The former name for a group of gram-positive bacteria, some of which form branching filaments that resemble mycelium of fungi. The name actinomycetes has now been replaced by "actinobacteria."

Adventitious bud—A bud that can arise from a leaf, a stem section between nodes, or from roots, rather than from the leaf axil.

Alkaline soils—Having a pH (a measure of hydrogen ion concentration) that is above pH 7.0; also referred to as higher pH or sweet soils.

Alternate host—Used with reference to some of the rust fungi: an alternate host is a second species of plant that hosts the rust fungus for a portion of its life cycle.

Anamorph—The imperfect or asexual stage of a fungus.

Angular leaf spot—A spot with straight sides, bounded by the leaf veins.

Anthracnose diseases—Certain diseases caused by *Colletotrichum, Discula, Gloeosporium* and a few other genera of fungi, typified by leaf spots and blights that spread easily by spore splash.

Antibiotic—A chemical, produced by a microbe, which is toxic to another microbe.

Aphids—Soft-bodied insects with piercing-sucking mouthparts, classified in the order Hemiptera. Some aphids vector (spread) viruses.

B

Bacteria (sing. bacterium)—One-celled microorganisms that cannot be seen with the naked eye; certain ones can cause plant diseases.

Bactericide—A pesticide that kills bacteria.

Basal plate—The flattened stem tissue at the base of a flower bulb from which scales, flower stalk and roots are derived. Damage to this tissue during shipment or handling will result in poor garden performance.

Biological control—Control of pests or diseases by using predators, parasites, or disease-causing organisms instead of using chemical pesticides.

Blight—Extensive tissue death that results when conditions are opportune for leaf spotting fungi and bacteria. Infection occurs throughout the plant and is not localized.

Broad mite—A tiny tarsonemid mite that feeds on the undersides of leaves or within the leaf buds of plants, leading to leaf distortion.

Bulbs—Modified shoots that provide food storage for dormant plants. Basal plates of bulbs give rise to root systems after planting.

C

Calyx—Grouped sepals at the base of a flower.

Canker—A dead area on a branch or stem that often causes wilting or dieback of the branch or plant beyond that point.

Chemical control—Control method based on use of chemical pesticides or disinfestants to reduce or eliminate a pest population.

Chlamydospores—Resistant spore stages formed by some plant-parasitic fungi.

Chlorosis—The yellowing of a plant's normally green tissue, due to the absence of chlorophyll.

Colonies—Circular patches of plant or fungal growth.

Color breaking—Interruption in the normal color of petals or leaves with either darker or lighter areas.

Conidia—Asexual spores of certain fungi.

Conidiophores—Microscopic stalks bearing conidia (asexual spores) of certain fungi.

Contagious—Capable of spreading; plant diseases caused by pathogens are all contagious to some degree.

Corms—Propagative units of gladiolus and some other herbaceous perennials composed of stem tissue modified for food storage.

Cormels—Newly produced, small corms.

Cortex—A cylinder of plant cells between the outer skin of the stem or root (epidermis) and the inner conductive (vascular) tissue (stele).

Crown gall—A common name for the plant disease caused by the bacterium *Agrobacterium tumefaciens*. The disease is typified by a gall at the stem base (root crown) of the plant.

Crown rot—Rot of a plant near the soil line, the "root crown".

Cultural control—Methods that do not involve use of chemical or biological techniques but result in pest or disease control. Example: Sanitation.

Culture indexing—A systematic program used by specialist propagators of some major ornamental plants to check for the presence of systemic bacteria and fungi contaminating stock plants before propagating the plant material for distribution.

Cyclamen mites—Tiny tarsonemid mites that feed in flower and leaf buds, causing distortion.

D

Damping-off—Seedling collapse due to rot of seeds before or after germination, generally due to fungal infection.

Desiccation—Drying of the plant tissue due to lack of water.

Dieback—The progressive death of the shoot or branch tip; caused by a root or stem disease, insect injury, nematode injury or cold weather, to name a few.

Disease—Any disturbance of a plant that persistently interferes with its normal structure, function, or economic value. (Can be caused by certain fungi, bacteria, viruses, nematodes, parasitic plants, and a few other types of organisms).

Disinfectant—An agent that kills or inactivates pathogenic organisms within the plant or plant part.

Disinfestant—An agent that kills or inactivates organisms present on the surface of the plant or plant part, tools, potting bench or other structure associated with the plant.

Division—Method of vegetative propagation for many perennials; often improves vigor of the transplants.

DNA (Deoxyribonucleic acid)—The principal, complex organic acid of the chromosome, representing the hereditary material of the cell.

Dodder—a type of parasitic plant

Dormant—Not in an active stage of growth. Plants are mostly dormant in winter, but some herbaceous perennials exhibit summer dormancy as well.

Downy mildew—A common name for a group of foliar plant diseases characterized by the "downy" growth of the causal organism on the underside of leaves.

E

Epidemic—An outbreak of a disease in a high percentage of the population of the plant.

Epidermis—The outer skin of the leaf, stem or root.

Eradicant—A pesticide that can kill a target pathogen or pest after it has infected a plant.

F

Foliar nematodes—Nematodes of the genus *Aphelenchoides* that infest the soil and will move up onto foliage when it is wet to infect plants through the stomata, causing leaf spots.

Fumigation—The action of disinfesting the soil, structures or growing surfaces, usually through use of a volatile gas or liquid pesticide.

Fungi (sing. fungus, adj. fungal)—Organisms (often microscopic) lacking chlorophyll, some of which cause rots, molds, and other plant diseases.

Fungicide—A pesticide that kills fungi or stops their activity.

G

Gall—Swollen area on leaves, stems and roots caused by a variety of pests, including certain insects, mites, fungi, bacteria and nematodes.

H

Hardy—Capable of surviving a hard frost; said of truly perennial plants in northern U.S. gardens.

Herbicide—A pesticide used to kill plants or inhibit their growth.

Host—The living plant or animal that a pest depends upon for survival during at least part of its life cycle.

Hybrids—The result of crosses between two plant species, usually within the same genus.

Hypocotyl—The part of a seedling above the radicle (which will become the root) and below the cotyledons.

I

Immune—A state of not being affected by a disease or poison (total resistance).

Infection—The establishment of a pathogen or parasite within the host, resulting in disease.

Infestation—Pests that are found in an area or location where they are not wanted.

Inoculum—Potential infective units of a pathogen, including spores and bacterial cells.

Insecticide—A pesticide used to control insects.

Intergeneric hybrids—The results of crosses between plant species belonging to two different genera.

Integrated Pest Management (IPM)—The systematic use of multiple approaches or methods of pest control; includes cultural, biological, mechanical, and chemical practices.

Invasive exotic plant—A plant with origins in another country or continent that grows and spreads too aggressively, thus becoming a noxious weed.

L

Lacebug—Insects in the Homoptera (*Stephanitis* spp.) with piercing-sucking mouthparts, usually feeding on the undersides of leaves, often producing conspicuous droppings.

Leaf abscission—Leaf drop.

Leafhopper—Insects in the Homoptera with piercing-sucking mouthparts, in some cases able to spread phytoplasmal, bacterial or viral diseases.

Leafminer—Larvae of certain species of flies, moths, or beetles, leafminers tunnel inside leaves as they feed, producing either discolored blotches or serpentine mines.

Line patterns—Zigzag or other patterned lines in yellow, white, brown or black on an otherwise green leaf; generally caused by virus infection.

M

Microsclerotia—Tiny dark bodies (masses of hyphae with a thick rind) formed by certain fungi as survival structures.

Mites—Tiny animals related to insects; they have eight jointed legs, two body regions, and no antennae (feelers) or wings.

Mosaic—A pattern of light and dark areas in a leaf that is frequently a sign of a virus infection.

Mottle—A symptom of irregular light and dark colored areas on plant parts that is frequently caused by a virus.

Mulch—A layer (usually 2-3 inches) of some organic material used to surround ornamental garden plants in order to reduce watering demands and discourage weed growth.

Multiseptate—Having strands that are divided into a chain of individual cells, said of a fungus.

Mycelium—The vegetative part of a fungus that corresponds to leaves, stems and roots of a plant. It is composed of a network of tiny strands.

N

Necrosis—Death of cells; in plants this usually results in the affected tissue turning brown.

Necrotic—Dead.

Nematode—Non-segmented roundworms; the plant-parasitic species feed on or in plants and are generally microscopic in size.

O

Oedema—Blisters on the leaf undersurface caused by physiological stress and excess water in the tissue.

Overfertilization—Providing more fertilizer to plants than they need for attractive growth, particularly excess nitrogen—thus creating overly succulent or leggy growth, favoring certain diseases, and potentially harming the environment.

P

Parasite—An organism that lives and feeds in or on another plant or animal (known as the host) and obtains all or part of its nutrients from it.

Pasteurization—The destruction of selected pests in soil or other growing media by use of heat or chemicals.

Pathogen—A disease-causing organism.

Pedicel—The slender stalk of a flower, or the stem of a single flower within a cluster of flowers.

Pest—An unwanted organism (insect, mite, nematode, bacterium, fungus, virus, weed, etc.) that attacks food, fiber or ornamental plants.

Petiole—The structure attaching a leaf to the stem.

pH—A measurement scale used to express the acidity or alkalinity of a solution. Examples: a pH of 7, the value for pure distilled water, is regarded as neutral; a pH of 1 to 7 is in the acid range; and a pH of 7 to 14 is in the alkaline range.

Phloem—The part of the conductive (vascular) system in a plant composed of cells responsible for food transport within the plant.

Phyllody—A symptom caused by phytoplasmas in which some ray and disk florets develop abnormally, forming leaves instead of flower parts

Phytoplasma—An organism similar to viruses in the symptoms it produces in plants, but more like a bacterium in its biology.

Phytotoxic—Harmful (poisonous or injurious) to plant life.

Pith—The central tissue in the stem.

Plasmid—A circular piece of DNA that is self-reproducing but not part of the hereditary makeup of the host or plant.

Potexvirus group—*Potato virus X* and its close relatives, all with flexuous rod-shaped virus particles.

Powdery mildew—A common name for a group of diseases on foliage, characterized by a powdery white or gray growth of the causal fungus.

Protectant—A pesticide applied to the plant surface before infection by a pathogen.

Pubescent—Covered with short hairs.

Pustule—Blister-like fruiting structure of a rust fungus, usually found on the leaf undersurface. When the structure ruptures, the colored reproductive spores are revealed.

Pycnidia (sing. **pycnidium**)—Spore containers that are the fruiting structures of some fungi, visible to the naked eye as tiny dots or bumps on the plant surface.

R

Resistant plant—One that is genetically less susceptible than others to a disease.

Rhizome—A modified stem that grows horizontally underground, producing both roots and shoots at nodes.

Rhizomorphs—Dark, cable-like structures containing fungal strands with a tough, drought-resistant outer covering, used for survival and dispersal of some fungi including *Armillaria mellea*, the shoestring root rot fungus.

Ringspot—A spot formed by concentric rings of tissue that are colored differently than the normal tissue (often yellow or brown); usually caused by a virus infection.

RNA (Ribonucleic acid)—A component of the cell sap and nucleus of plants and animal cells; representing a small portion of the chromosomes of all cells. The genetic material of many plant viruses is RNA.

Root knot nematode—A root-feeding nematode that produces galls on roots as it feeds within them, resulting in stunted plant growth both above and below ground.

Runners—Stolons of certain plants (e.g. strawberry), allowing vegetative spread.

Rust—A common name for a group of diseases caused by fungi; named for the colored (sometimes rusty brown) spores formed in pustules on infected plant parts.

S

Sanitation—Any activity aimed at eliminating or reducing the numbers of a pest present in a given area. Includes removal of diseased or infested plants or plant parts, weeds, and dead plant material.

Saprophyte—An organism that can live on dead or decaying organic matter.

Scapes—Flower stalks without attached leaves, emerging from the ground, as for daylilies

Sclerotia (sing. sclerotium)—Resistant, seed-like structures formed by some fungi to overwinter or remain dormant until conditions are favorable for growth and/or plant infection.

Scorch—A symptom of browning and drying along the edge of a leaf, or browning and drying of the ends of shoots, indicating acute water deprivation.

Senescence—Normal decline and death of plant tissues due to aging.

Shoot proliferation—Disturbance of the normal plant growth regulation, such that apical dominance is lost and many, usually small and distorted, shoots are produced from a node, often on or near the base of a stem; sometimes caused by plant pathogens.

Slugs—Molluscs without shells that are voracious foliage pests in gardens, especially in wet climates.

Smuts—Diseases caused by a group of fungi closely related to the rusts, many of them causing symptoms on flower parts or leaves.

Soft rot—A disintegration of the normally organized cell structure of plant parts due to the action of certain bacteria or fungi, resulting in a wet, slimy, often smelly decay. Storage organs (e.g. corms, tubers) are especially susceptible.

Soilborne—Inoculum occurring on or in the soil, often referring to pathogens that may persist as saprophytes on decaying organic matter.

Solarization—A non-chemical means to kill pathogens, insects and weeds in soil under a clear plastic tarp, using the heat generated by sunlight, over 4 to 6 weeks.

Spittlebug—An insect whose larval stage forms conspicuous blobs of what appears as "spittle" to serve as protection while it feeds on stems of plants.

Sporangiophores—Microscopic stalks bearing spores (called sporangia) of certain pathogens, including *Phytophthora* spp. and some downy mildews.

Spore—A fungus structure, analogous to the seed of a plant, which serves to reproduce and spread the fungus.

Sporulation—Spore production by a fungus, which serves for reproduction and dissemination.

Stele—The conductive tissues of roots, composed of phloem and xylem cells.

Stem and bulb nematode—A nematode (*Ditylenchus dipsaci*) affecting primarily bulbs, causing necrosis and distortion of the above-ground parts of plants that it infects.

Stolons—Horizontal stems that grow either on the soil surface or below it, having long internodes and producing new shoots at their tips.

Sunscald—Irregular patches of discolored and dry tissue on plants due to excessive sun exposure, most common on leaves of variegated varieties.

Susceptible plant—Capable of being injured or diseased by a pest; not resistant or immune.

Symptom—Something abnormal about a plant's appearance.

Systemic—Something taken up by one part of a plant and moved to another part, distributed via the vascular system. The term refers to certain plant diseases as well as to certain pesticides.

T

Telial, Teliospore—Referring to one of the life stages of a rust fungus, often the stage that overwinters.

Thrips—A tiny, hard-to-see winged insect (order Thysanoptera) with piercing mouthparts, commonly vectoring both *Tomato spotted wilt virus* (TSWV) and *Impatiens necrotic spot virus* (INSV).

Tolerant—The state of being only mildly affected by a disease (partial resistance).

Toxicity—How poisonous a pesticide is to an organism; the ability of a pesticide to produce injury.

Tubercles—On the microscopic oval eggs of broad mite, ornamentations appearing as rows of whitish pegs—a key diagnostic feature to distinguish these eggs from those of cyclamen mite.

Tubers—Large, swollen underground stem pieces used for propagation of some herbaceous perennials.

V

Varieties—Selections of plant species with unique traits that are maintained by vegetative propagation; for ornamentals, these are usually variations in flowering, or sometimes foliage variegation or plant size.

Vascular bundles—Conductive tissue (xylem and phloem) arranged in strands within plants.

Vascular infection—Systemic infection of the xylem or phloem of a plant by a fungus, bacterium or other pathogen, which can result in decreased water transport, wilting and stunting.

Vascular tissue—The conductive tissues of a plant, composed of phloem and xylem cells and known as the stele in the root system.

Vector—A carrier of a disease-producing organism (pathogen); an insect or other animal that transmits a pathogen.

Virescent—With abnormal green color, such as that seen in flowers as a symptom of aster yellows disease.

Viroid—A virus-like submicroscopic pathogen with genetic material (RNA) that is not encased in a protein coat.

Virus—A submicroscopic pathogen that needs living cells to grow; some can cause disease in plants. Viruses are too small to be seen with a regular microscope, so an electron microscope is needed.

Virus indexing—A systematic program used by specialist propagators of major ornamental plants to check for the presence of known viruses in stock plants before propagating this plant material for distribution.

W

Water molds—Organisms classified as Oomycetes, once considered fungi, now known to be more closely related to algae (e.g. *Pythium* and *Phytophthora* species and downy mildews).

Water-soaking—Dark, greasy or wet-looking tissue surrounding leaf spots caused by bacteria and sometimes fungi.

White blister—A common name for the disease caused by *Albugo tragopogonis*, a water mold, on its hosts.

White rust—A common name used for two very different diseases: 1) the true rust disease caused by *Puccinia horiana* on chrysanthemum and 2) the leaf disease (also known as white blister) caused by *Albugo tragopogonis* on its various hosts

White smuts—Diseases caused by fungi in the genus *Entyloma*, which cause foliage and stem symptoms.

Witches' broom—A symptom that appears as a cluster of small, weak shoots emerging from the same point on a stem or branch; caused by bacteria, fungi, phytoplasmas, mites and other pests.

X

Xylem—Part of a plant's conductive (vascular) system, composed of cells responsible for water transport from the roots to the leaves.

Index

A

B

C

D

E

F

G

H

M

N

O

P

Q

R

S

X

Y

Z